Hydrothermal Behavior of Fiber- and Nanomaterial-Reinforced Polymer Composites

Hydrothermal Behavior of Fiber- and Nanomaterial-Reinforced Polymer Composites

Ramesh Kumar Nayak

Bankim Chandra Ray

Dibyaranjan Rout

Kishore Kumar Mahato

CRC Press

Taylor & Francis Group

Boca Raton London New York

CRC Press is an imprint of the
Taylor & Francis Group, an **informa** business

CRC Press
Taylor & Francis Group
6000 Broken Sound Parkway NW, Suite 300
Boca Raton, FL 33487-2742

and by

CRC Press
2 Park Square, Milton Park, Abingdon,
Oxon OX14 4RN

First issued in paperback 2021

© 2020 by Taylor & Francis Group, LLC
CRC Press is an imprint of Taylor & Francis Group, an Informa business

No claim to original U.S. Government works

ISBN 13: 978-0-367-25442-1 (hbk)
ISBN 13: 978-1-03-224085-5 (pbk)

**Visit the Taylor & Francis Web site at
http://www.taylorandfrancis.com**

**and the CRC Press Web site at
http://www.crcpress.com**

Publisher's Note
The publisher has gone to great lengths to ensure the quality of this reprint but points out that some imperfections in the original copies may be apparent.

Contents

Preface

The trending technology in materials science is based on the development of a new class of materials for different engineering applications. Fiber reinforced polymer (FRP) composites have been developed and are used in different sectors, such as automotive, aerospace, marine, civil infrastructure, household appliances, and so on. The extensive acceptance of these materials is due to structural tailorability, light weight, and cost-effective manufacturing processes. Advancements in these materials have significantly contributed toward their potential exploitation in high performance and high precision mobile and immobile structural applications. The most attractive strength of the composite materials lies in their superior specific properties (e.g., strength-to-weight ratio and modulus-to-weight ratio) in conjunction with good impact strength, corrosion resistance, and fatigue and damping characteristics, which motivates engineers to use these materials in quite a wide spectrum of diversified applications. The interface strength between the fiber and matrix is the heart of the FRP composites. Aging in different environmental conditions is essentially fading away their durability and reliability. The environmental parameters may include temperature, moisture, ultraviolet light, and other high energy radiation (electromagnetic, microwave, γ rays, etc.). These parameters play important roles in altering the physicochemical structure of the polymeric material. Therefore, the durability of the nano-composites is a challenge in hydrothermal environments. Scientists and researchers have addressed these issues in a different forum. In this book, an attempt has been made to illustrate the durability of these FRP composites at various sets of environmental parameters. Nevertheless, it is focused on the effect of different nanoparticles or fibers on the durability of the composites in hydrothermal environments. The book has taken the inference of the work performed by different researchers and scientists in the development and characterizations of nano-composites and the effect of different environments on their durability. The ingression of water into the composites is a very common phenomena in the hydrothermal environment, and, hence, the authors have emphasized the effect of a hydrothermal condition on the performance of nano-composites. The first chapter has been focused on the different types of nano-composites and their application in different fields. The common fabrication methods and characterization of nano-composites have been explained in Chapter 2. Organic and inorganic nanofillers are added into the FRP composites to enhance their properties. Nanofillers are added individually or combinedly into the FRP composites to improve the desired properties. Both natural and synthetic fibers are used in the development of composites, and nanofillers are added to the fibers to improve their mechanical properties and durability. However, these composites are subjected to a hydrothermal environment during service. The diffusion of water into the nano-composites for different fibers is different. Hence, the water diffusion models developed by different scientists and researchers have been discussed in Chapter 4. Furthermore, the effect of a hydrothermal environment on the mechanical properties of the nano-composites has been discussed in Chapters 5–8. The authors have conceived and compiled the hydrothermal damage and degradation of the advanced structural FRP nano-composites and highlighted the current cutting edge research involving the addition of nanofillers from it. This book is evolving to provide a platform to the researchers and engineers who are working on FRP composite materials. It will help them to design their composites for better durability in hydrothermal environments.

The authors would like to take this opportunity to extend their heartfelt gratitude to the Maulana Azad National Institute of Technology, Bhopal, National Institute of Technology, Rourkela, and KIIT University, Bhubaneswar, India and the beautiful people associated with it.

Authors

Ramesh Kumar Nayak is working as an assistant professor in the Department of Materials and Metallurgical Engineering, Maulana Azad National Institute of Technology, Bhopal, India. He graduated from the Indian Institute of Technology, Kanpur, India in 2005 and received his PhD from KIIT University, Bhubaneswar (awarded as Institution of Eminence (IOE) by the Government of India) in 2016. He has sound exposure to similar industries and has worked in reputed organizations such as General Motors India Technical Center, Bangalore, Hindustan Aeronautics Limited, Bangalore, and DENSO International India Pvt. Limited, Gurgaon in different capacities.

He has been working in the development of composite materials and casting technology areas for the last 14 years. He teaches polymer engineering, solidification and casting, composite materials, and corrosion engineering to undergraduate and graduate scholars. He has guided one PhD and several MTech scholars in the area of composite materials. He has published his work in peer-reviewed international journals and contributed to the world literature in the field of materials science. He has filed for two patents on the development of new class of composite materials. His passion is to work closely with industrial problems and develop new technology/process for the benefit of society.

Bankim Chandra Ray is a dedicated academician with more than three decades of experience. Bankim Chandra Ray currently holds a full professor position at the National Institute of Technology, Rourkela, India. He was awarded a PhD from the Indian Institute of Technology, Kharagpur, India in 1993. Apart from instructing students in the field of phase transformation and heat treatment, he has also guided several PhD scholars. He has made a seminal contribution in the field of phase transformation and heat treatment and composite materials.

An adept administrator, he has also served as the incumbent dean of Faculty Welfare, head of the Department of the Metallurgical and Materials Engineering, and coordinator of Steel Research Center at NIT Rourkela.

His research interests are mainly focused on the mechanical behavior of FRP composites. He is leading the Composite Materials Group at NIT Rourkela, a group dedicated to realizing the technical tangibility of FRP composites (https://www.frpclabnitrkl.com). With numerous highly cited publications in prominent international journals, he has contributed extensively to the world literature in the field of material science. He also holds a patent deriving from his research. He has collaborated actively with several coveted societies, such as the Indian Institute of Metals and

Indian National Academy of Engineering. As an advisor to the New Materials Business, Tata Steel Ltd., he has been instrumental in facilitating the steel honcho's foray into the FRP composites business.

Dibyaranjan Rout is currently working as an associate professor at the School of Applied Sciences (Physics), Kalinga Institute of Industrial Technology (KIIT) Deemed to be University. He has been awarded a PhD degree in physics from the Indian Institute of Technology Madras, Chennai-36 in 2006. Following his PhD, he worked as a research assistant professor at the Korea Advanced Institute of Science and Technology, South Korea from 2006 to 2011. Apart from teaching UG/PG students, he has guided several PhD and master students. His research interests lie in the frontier fields of functional materials (lead-free piezoelectric and multiferroic materials), nanoparticle synthesis, and polymer nano-composites for energy and environmental applications. To his credit, he has published 50 research articles in international journals of high impact and presented more than 100 papers in international/national conferences. He has active collaborations with researchers of several premier institutes in India and Abroad.

Kishore Kumar Mahato has been working as an assistant professor at Vellore Institute of Technology, Vellore, India since June 2019. He has completed his PhD from the National Institute of Technology, Rourkela, India on the topic "Environmental durability of multiscale glass fiber/epoxy composites: An assessment on mechanical properties and microstructural evaluation" under the supervision of professor Bankim Chandra Ray. He has published around 20 research articles in different SCI and Scopus indexed journals and 2 book chapters. His current research work is focused on the failure and fracture behavior of fiber reinforced polymeric composites in different harsh environments. His investigations are focused on the assessment of the mechanical behavior of environmentally conditioned FRP composites through experimental and numerical analysis. Primarily, the polymer matrix and the existing fiber/polymer interface are susceptible to harsh and hostile in-service environments, which can alter the durability and integrity of fibrous polymeric composites.

1

Introduction

1.1 Polymer Nanocomposites

Polymer nanocomposites (PNCs), an appropriate synonym for nanoparticles in the form of rods, spheres, or sheets dispersed within the polymer matrix, has aroused tremendous interest, both in academia and industry since their early reports in the 1980s. The unique combination of nanoscopic fillers with a polymeric matrix offers a wide range of refined properties including increased strength, high moduli, heat resistance, decreased flammability, and gas permeability, which increase the biodegradability of biodegradable composites. These advantages result primarily from the reduction in the filler size to nanoscale, which increases the surface area to volume ratio and largely improves the interaction of the dispersed phases. Replacing the traditional polymer composites (with micro-sized fillers), PNCs have found their place in numerous applications like the automotive sector, packaging, sporting, aerospace, marine, medical, and so on [1,2]. The global polymer nanocomposite market is expected to register significant growth in the coming years and is expected to reach $66,876 million by 2022 with a growth rate of 3.7%. Thus, maintaining the scientific temper and global consumption of PNCs, the current chapter classifies the types of polymer nanocomposites and also describes their importance in terms of potential applications.

1.2 Classification

1.2.1 Introduction

Composites can be defined as an ensemble of two phases, blended/mixed to obtain the desired properties, tailor-made for specific applications. Nanocomposites constitute a subgroup of this bigger domain of multifunctional materials having at least one of the components in the nanoscale range. The arena of nanocomposites with unusual property combinations and design possibilities at a very low concentration of fillers has gained a special status in recent years owing to a high matrix to filler interfacial area (the so-called "Nano effect") and greater aspect ratio. The best example appears in Mother Nature in the form of bones, shells, and wood. Based on the matrix type, the nanocomposites are classified as ceramic matrix nanocomposite, metal matrix nanocomposite, and polymer matrix nanocomposite. A schematic diagram demonstrating the overall classification of nanocomposites is provided in Figure 1.1. Among these categories, PNCs have reached a phenomenal status in industrial and real-life applications due to their ease of production, lightweight, ductile nature, high strength, better resistance to corrosion, fire, and acids, higher fatigue strength, and much more. Polymer matrices can be broadly divided into two major groups based on their response to heat: (i) thermosetting polymers and (ii) thermoplastic polymers, and are discussed in the following sections.

1.2.2 Thermosetting Polymers

Thermosets are the polymers that strengthen when heated, but after the initial forming, they cannot be heated or remolded. Those are cross-linked structures that harden primarily by undergoing an irreversible chemical curing process which is very much similar to a hard-boiled egg that cannot disfigure by additional heating once heated. The three-dimensional network or the cross-linkages provide mechanical strength and greater resistance to heat degradation and chemical attack. A unique coalescence of properties like chemical resistance, thermal stability, and structural integrity in thermoset polymers propose their candidature in a wide range of applications including automotive, electrical appliances, and lighting. Additionally, the thermosets can offer a number of benefits as compared to their metal counterparts: (i) available molded-in tolerances, (ii) choice of color and surface finishes, (iii) outstanding dielectric strength, (iv) high strength-to-weight ratio and performance, (v) lower tooling/set-up costs, (vi) low thermal conductivity and microwave transparency, (vii) reduced production costs over fabrication using metals, (viii) resistance to corrosion effects and water, etc. The commonly used thermosets are phenolics, epoxies, backalite, polyurethanes, polyester, melamine formaldehyde, polyvinylidene fluoride (PVDF), silicone, etc. A representative figure of the polymerization of some of the thermosetting polymers is given in Figure 1.2 [3,4]. In spite of so many advantages, those polymers are associated with a few drawbacks such as brittleness, awful surface finish, poor thermal conductivity, and unsuitability of the polymer for vibration applications due to a higher rigidity.

1.2.3 Thermoplastic Polymers

Thermoplastics are a class of polymers that can be repeatedly softened and reformed upon enforcing pressure and heat. This is because the thermoplastics are characterized by physically ordered crystalline domains that are

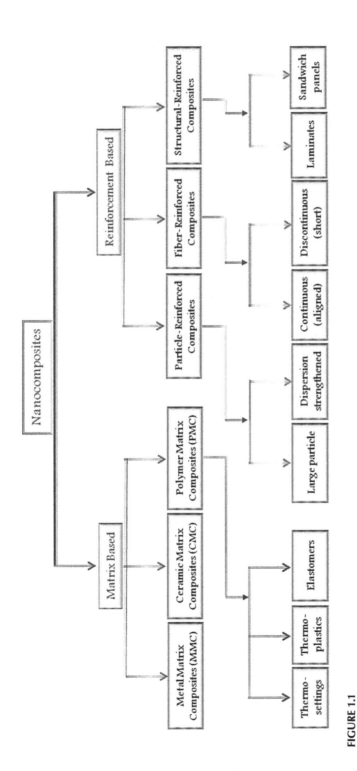

FIGURE 1.1
A schematic diagram displaying the overall classification of nanocomposites.

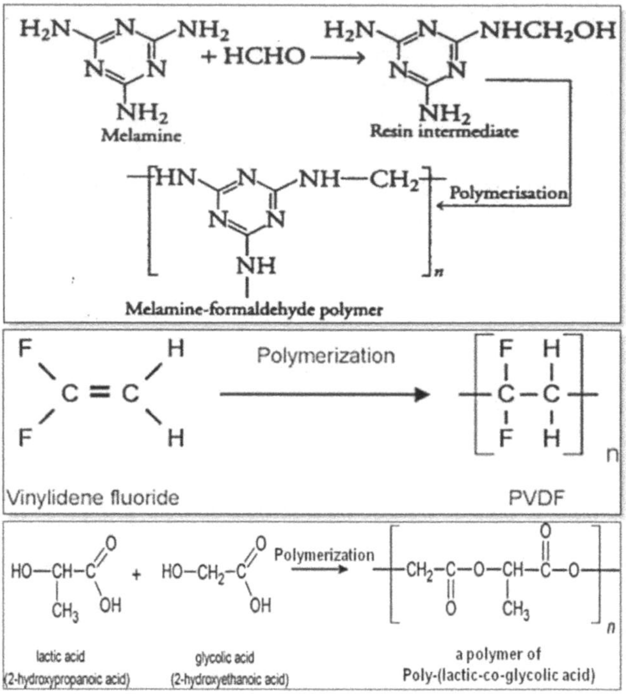

FIGURE 1.2
Polymerization of some thermosetting polymers. (From Hartono, A. et al. Preparation of PVDF film using deep coating method for biosensor transducer applied, *2013 3rd International Conference on Instrumentation, Communications, Information Technology and Biomedical Engineering (ICICI-BME)*, 408–411, 2013.)

not chemically cross-linked. Such an arrangement allows flexibility in the reshaping/remolding of the polymers on the application of heat, even after initial forming. Another remarkable feature of thermoplastics is the desired physicochemical properties of the resulting composite which can be well-tuned by permuting the chemical content of the chain and also the length of the chain. Some flexible and lightweight polymers such as polystyrene and polymethacrylate (PMMA) prove to be excellent alternatives to glass substrates and metals which cannot handle bends. Here, some of the general advantages of thermoplastics are summed up:

- Low Density
- Corrosion Resistance
- Excellent Mechanical Strength
- Easy Fabrication of Flexible Films
- Low Cost
- Toughness and Ductility

Other commonly used thermoplastic polymers are polyether ether ketone, poly phenylene, poly ethylene, polycarbonate, polysulfone, polyvinyl chloride and nylon, etc. Polyvinyl chloride is often used to make pipes, while polyethylene gas tanks are used in residential and commercial applications to transport natural gas. Similarly, polyamide is used for the production of ropes and belts. Although thermoplastics offer a vast scope for applications, yet they are susceptible to creep when stretched under exposure to long term stress loads. The monomer structure of some of these polymers is given in Figure 1.3 [5]. An effective approach to address the drawbacks of the polymer matrices is to reinforce them with fillers, particularly on the nanoscale in the form of spheres, rods, or sheets. Nano-reinforcements include mainly polymer, metal, or ceramic fillers.

1.2.4 Polymer-Polymer

Polymers have been attracting the attention of researchers, not only as matrix material, but also as nanofillers owing to their unique mechanical, electrical, and chemical properties. However, research in this field is still in its infancy and is expected to bring in many fruitful results. Chitosan is a well-known and abundantly available organic polymer which is biocompatible in nature. Chitosan when used as nanofiller in epoxy resin proves to be an effective anticorrosion coating and can be applied to mild steel substrates under ambient conditions [6]. Similarly, the conducting polyaniline/polyvinyl alcohol nanocomposite blend, prepared by chemical oxidative synthesis, could change their electrical resistance with humidity levels

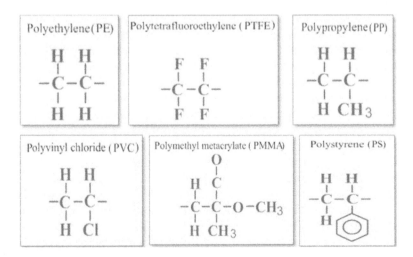

FIGURE 1.3
Monomer structure of different thermoplastic polymers. (From https://ask.learncbse.in/t/how-is-the-following-resin-intermediate-prepared-and-which-polymer-is-formed-by-this-monomer-unit/10687.)

(acting as humidity sensors) [7]. An extensive review of the sensing applications of polyaniline/polymers such as polyvinyl acetylene, PMMA, and polyvinyl alcohol (PVA) is provided by Sen et al. [8]. Even the electrospun nanofiber mat of polyaniline (PANI) and polyethylene oxide acted as better NH_3 sensors with a short response time of 6 sec as compared to the conventional film. The electrical conductivity of the nanocomposite mat is about two orders higher than that observed in PANI/polyethylene oxide films with a much lower polyethylene oxide concentration. Other than sensing properties, PANI-based nanocomposites are also known for their mechanical strength (tensile), ultrasonic attenuation, and thermal conductivity when prepared by different chemical routes. Guo et al. [9] prepared PANI nanoparticles by oxidative polymerization and incorporated them into an epoxy matrix by three different approaches. The electrical conductivity and tensile strength increased significantly at 8 wt% PANI loading as compared to previously reported 10 wt% PANI loading. Wan et al. [10] could ensure a better dispersion of PANI nanoparticles in the epoxy matrix by ultrasonication with mild ultrasonic waves and in turn, improved the thermal conductivity and ultrasonic attenuation behavior of the nanocomposites than virgin epoxy.

1.2.5 Polymer-Metal

Nanoscopic metals possess a considerable property change upon size reduction as compared to their bulk counterparts. The special and interesting properties include biocidal activity, catalytic action, ferromagnetism and superparamagnetism, chromatism (in metals like gold, silver, copper, etc.), and photo- and thermoluminescence produced by electron confinement and surface effect (quantum-size effects). Besides, these nano-sized metals become thermal and electrical insulators due to the disappearance of band structure, extremely chemically reactive, and also show a different set of thermodynamic parameters (e.g., own very low melting temperatures). However, these advanced materials showing extraordinary behavior are associated with some limitations. Most of the nano-metals are very unstable and can form aggregates due to a higher surface energy and be easily oxidized by moisture, air, SO_2, etc. So, by maintaining the functionalities of these metals intact, they can be embedded as fillers into a polymer matrix, and the resulting composites exhibit some riveting properties starting from particle plasmon resonances producing characteristic optical behavior to giant magnetoresistance and a magnetic property governed by a single ferromagnetic domain. The polymer-metal nanocomposites were established since long ago in 1835, when a gold salt was reduced in aqueous solution in the presence of gum arabic, and eventually a nanocomposite was obtained in the form of a purple solid by co-precipitation with ethanol. This has acted as a stepping stone and with the progress of technology; numerous efforts have been made toward the preparation and utility of these composites. A hybrid nanocomposite with silver nanoparticles dispersed in a conducting polymer, i.e., poly(o-toluidine) (a derivative of polyaniline) was prepared by in-situ chemical oxidative polymerization [11]. The incorporation of conductive Ag particles into the polymer matrix improves the conductivity of the polymers and was expected to find their utility as biosensing materials. Silver and copper nanoparticles are a class of materials which are known for their biocidal (antibacterial, antiviral, and antifungal) nature. Super absorbent hydrogel nanocomposites prepared by impregnating silver nanoparticles into polyvinyl alcohol and sodium alginate exhibit good antibacterial activity on gram-positive and gram-negative microorganisms. Similarly, coatings based on polymer/copper nanocomposites also proved to be potential antibacterial materials and were extensively used as marine antifouling coatings. In another study, air-stable iron nanoparticles encapsulated with poly (vinyl pyrrolidone) nanofibrous membranes proved to be potential candidates for catalytic systems and groundwater purification [12].

1.2.6 Polymer-Ceramic

Dielectric nanoceramics encompassed with a high dielectric constant are found to be critical for a wide range of applications such as bypass capacitors, microelectronics, energy storage devices, pulsed power systems, systems on package technologies, etc. Apart from this, the ferroelectric class of ceramics acts as prospective material for sustainable energy harvesting. However, the brittleness and low dielectric strength are the major hindrances from the application point of view. Ceramics as nanofillers when introduced into the flexible and lightweight polymer matrix combine the advantage of both ceramic and polymer giving rise to outstanding functionalities like giant permittivity, low loss, high dielectric breakdown strength, and long life cycle [13,14]. Some of them are ferroelectric like $BaTiO_3$, $BaSrTiO_3$, or non-ferroelectric fillers (boron nitride, graphene oxide). In a study, TiO_2 nanofibers embedded with $BaTiO_3$ nanoparticles could notably increase the dielectric constant in poly(vinylidene fluoride-hexafluoropropylene) P(VDF-HFP) [15]. In another study, Young's modulus and tensile strength of the barium sodium niobate-polystyrene nanocomposites show an extensive increment with the incorporation of barium sodium niobate filler [16].

1.2.7 Polymer-Carbon

Carbonaceous nanoparticles have been considered as the embryonic materials among the fillers in recent years due to their extraordinary thermal, mechanical, and electrical properties. Carbon exists in different allotropic forms such as diamond, graphite, fullerenes, single-walled and multi-walled carbon nanotubes (SWCNTs and MWCNTs), carbon nanofibers, and graphene (Figure 1.4). Graphene, obtained from the micromechanical cleavage of graphite is a layer of sp2 hybridized carbon atoms arranged in a honeycomb structure with a thickness of one atom. A single graphene

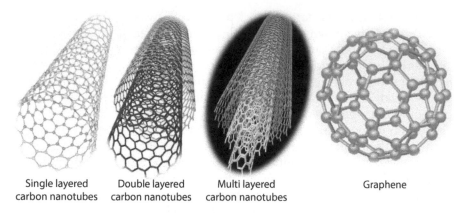

| Single layered carbon nanotubes | Double layered carbon nanotubes | Multi layered carbon nanotubes | Graphene |

FIGURE 1.4
Structure of selected allotropes of carbon.

sheet rolled up into a cylinder is termed an single-wall carbon nanotubes (SWCNT), while two or more concentric cylindrical sheets bounded by van der Waals forces are called an MWCNT. The length of the CNTs varies from several hundred nanometers to micrometers. Similar cylindrical nanostructures with larger diameters are known to be carbon nanofibers. On the other hand, fullerenes are hollow carbon spheres with pentagonal or hexagonal faces. These nanofillers with a high aspect ratio when reinforced into a polymer matrix yield superior mechanical strength, thermal and electrical conductivity, chemical inertness and stability, optical transparency, photoluminescence, biocompatibility, etc. [17]. The outstanding mechanical properties of nano-diamonds are utilized to enhance the mechanical strength of the polymers and biopolymers for structural applications with longer service life. It is reported that the incorporation of nano-diamonds to a final rubber compound increases its elasticity, elongation at rupture, abrasion resistance, and the tear resistance of vulcanized rubbers [18]. Graphene, also belonging to the same family of high strength and lightweight materials, is being highlighted for its use in the automotive sector since the reduction in weight of the automobiles increases their fuel efficiency. Remotely, Ford Motor Company has taken the initiative to use graphene parts in their vehicles (making a start with F-140 and Mustang) by the end of 2018 [19]. Graphene is also bestowed with high electrical conductivity (up to 6×10^3 S/cm), an amalgamation of which within lightly cross-linked polysilicon (e.g., silly putty), gives an ultra-sensitive electromechanical response which can be utilized for detecting a pulse, blood pressure, and even small spider steps. Carbon nanotubes possess promising strength, modulus, and conductivity. The tensile strength and Young's modulus of MWCNTs fall in the range 11–63 GPa and 270–950 GPa, respectively [20]. Due to its mechanical properties, NASA has introduced the concept of using low density and high strength CNTs as cost-effective space elevators.

1.2.8 Polymer-Natural Fiber

The use of natural fibers as nanofillers in polymer composites is fueled to over-come certain disadvantages of their synthetic or man-made counterparts such as carbon, polyethylene, aramid, glass, etc. Natural fibers are inherently high in specific strength and specific modulus, cheap, less dense, and easy to process; and most importantly, renewable and recyclable in nature. They can be sourced both from plants and animals. Commonly, the natural cellulose fibers obtained from plants can be categorized into bast fibers (jute, hemp, ramie, flax, kenaf), seed fibers (cotton, kapok, coir), leaf fibers (sisal, abaca, pineapple), grass and reed fibers (corn, rice, wheat), and core fibers (hemp, kenaf, jute), along with other kinds like wood and roots (Figure 1.5). A major part of concern behind the usage of natural fibers as nanofillers also arises from the depletion of unsustainable petroleum resources causing a greater amount of environmental pollution, the release of greenhouse gases, and moreover, a thirst for materials/products that can be recycled and reused. Certain regions of the European Union and Japan have already imposed the legislation stating that almost 80% of the automobiles manufactured should be recyclable. Such legislation may soon be globally enforced. So, inevitably, the manufacturers are trying to use products that are environmentally friendly and biodegradable. Companies such as Mercedes-Benz have taken the initiative of developing 50 polymer composites with natural fibers as nanofillers for its E class series of automobiles.

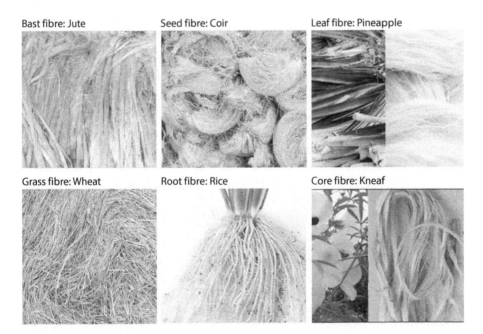

FIGURE 1.5
A list of natural fibers.

Moran et al. prepared poly (lactic acid)/cellulose nano-whiskers by twin-screw extrusion followed by injection molding [21]. The composites showed much better mechanical performance with a low filler content and were found to be suitable for packaging applications. Nano-celluloses are often used as reinforcing fillers in several polymer matrices due to their exciting intrinsic properties. Recently, nano-cellulose extracted from jute fibers, when included into a natural rubber matrix could cause a drastic improvement in the mechanical properties of the nanocomposites [22]. The tensile strength and Young's modulus of the material increased, while the elongation at break decreased. Further, in another interesting study by Salehpour et al. the biodegradability of PVA and cellulose nanofiber composites was investigated under composting conditions that were found to behave with no negative effect on plant growth [23].

1.3 Applications

1.3.1 Introduction

Polymer nanocomposites with their unprecedented property combinations and exceptional design possibilities are establishing themselves as high-performance materials of the twenty-first century and are used in multifarious cutting-edge technologies. A schematic diagram is provided in Figure 1.6 listing out various applications of polymer nanocomposites. A few of the applications are briefly discussed in this section.

1.3.2 Aerospace

Projecting heavy lift systems to the earth's lower atmosphere incurs a huge cost in terms of fuel prices. The fuel cost amounts to about 30% of the operational cost even in general aviation. Hence, there has been a continuous curiosity among the researchers and industries to replace the metallic components with lightweight materials, i.e., by employing polymer nanocomposites with superior features. Some materials like Fe-, Ni-, or Cr-coated nano-graphite reinforced into a polymer matrix were found to be suitable for microwave absorption [24]. A group from IIT Delhi, working on such coatings revealed that their microwave absorption frequency was in the range 300 MHz–1.5 GHz for Fe–Ni-coated nano-graphite [25]. Such nanoparticles when dispersed into a polyurethane matrix could file its candidature in defense aircraft as camouflage cover, nets, and other coated textiles for aerial surveillance. On a similar note, electrically conductive polyetherimide/carbon nanofiber/Ag-coated nanofiber composite film was prepared by Nair et al. [26]. These coatings were suitable for enhanced protection from lightning strikes and electromagnetic interference (EMI) shielding. However,

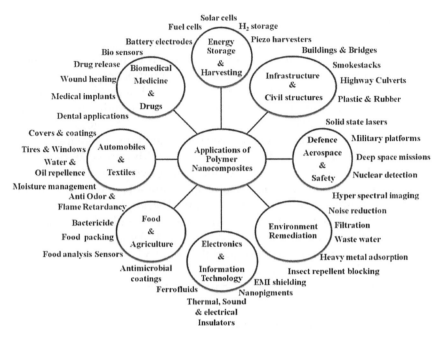

FIGURE 1.6
A schematic diagram listing various applications of polymer nanocomposites.

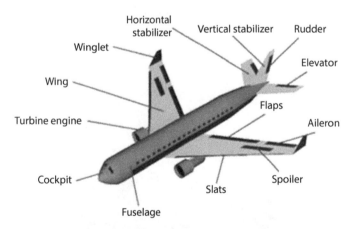

FIGURE 1.7
Polymer nanocomposites used in different parts of the Airbus A380. (From Mrazova, M., *Incas Bulletin.*, 5, 139, 2013. http://www.ewriteright.in/tutorial/writers6.php.)

among the various nanofillers used, CNTs are being considered to be progressive for their distinctive electrical and mechanical properties. This is quite evident from the implementation of polymer/CNT nanocomposites in a space shuttle and commercial aircraft such as Boeing 787 and Airbus A380 as shown in Figure 1.7 [27–29].

1.3.3 Automotive

With increasing global concerns for low fuel economy and low emissions in the case of land transportation systems, research is trending toward the low cost, high performance, and lightweight polymer nanocomposites. This class of novel materials is expected to increase the speed of production, environmental and thermal stability, and recyclability, while reducing the weight. The first initiative in this direction was taken by Toyota Motors in 1989 by introducing nylon-6/nano-clay composites commercially. Thereafter, General Motors in 2002 and Chevrolet in 2004 followed with nano-clay-based polymer composites for application in exterior automotive parts. Maserati engine bay covers were also made from nylon-6 and nano-clays which reduced its weight along with an increase in mechanical strength. Polyolefin matrix nanocomposites have also attracted kin attention recently due to their long-term durability and cold weather resistance [30,31]. This is exemplified by its use in the instrument panels of some of the prime players in the automotive sector like the 2000 Ford Focus and 2000 Pontiac Bonneville, door panels of the Mercedes-Benz E class, Porsche 986/996, and Honda Civic (Figure 1.8).

1.3.4 Infrastructures/Civil Structures

Polymer composites with nanofillers have always acted as game-changers about their use in structural components (buildings, bridges, and other engineered structures) which can be attributed to the high strength-to-weight

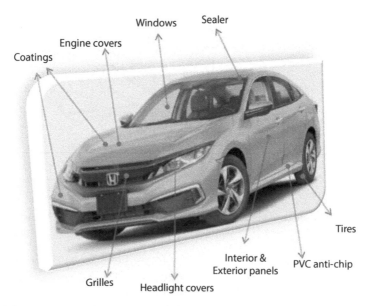

FIGURE 1.8
Lightweight automatic parts of a Honda Civic car.

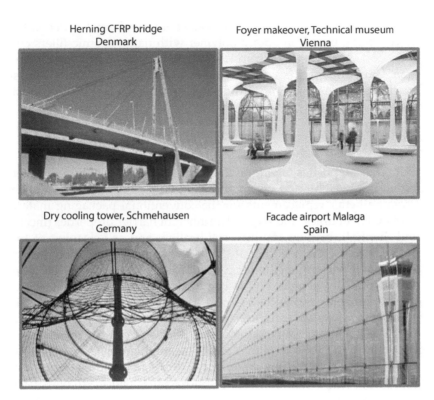

FIGURE 1.9
Polymer nanocomposites used in different civil structures. (From Moy, S., Advanced fiber-reinforced polymer (FRP) composites for civil engineering applications, in *Developments in Fiber-Reinforced Polymer (FRP) Composites for Civil Engineering*, Woodhead Publishing, pp. 177–204, 2013.)

ratio of the class of materials (Figure 1.9). They are also highly durable in terms of thermal, mechanical, and barrier properties [32,33]. One of the important components of civil structures is concrete. But a lot of improvements are expected in it concerning its increased durability, tensile strength, and reduced brittleness [33]. These issues can be well addressed by modifications with polymer nanocomposites. Acrylic acid-acrylamide, a copolymer with colloidal silica nanoparticles as fillers significantly reduces the porosity of the capillary system in concrete, while another composite of poly (vinyl alcohol) and montmorillonite could improve the strength properties of cement. Asphalt, another construction material, degrades with exposure to atmospheric oxygen and ultraviolet rays, and leads to its aging. This aging could be slowed down by blending with styrene-butadiene-styrene/organic-modified montmorillonite nanocomposites. Polymer composites with nano-fillers are also known for their inert behavior and, hence, frequently used as anticorrosive and barrier surface coatings to regress the effect of water and other ions. Composites with organo-clays are commonly used as barrier coatings to protect the civil structures against environmental aging and corrosion. A coating based on epoxy polymer with nano-ceramic fillers could shield the concrete structures from UV radiations, contamination, and deterioration. Gauvin et al. prepared molded nanocomposites from vinyl ester (VE) resin and nano-clays and found them durable against water diffusion and surface hydrolysis [34].

1.3.5 Food Packaging

Polymer nanocomposites, owing to their superior functionality, antibacterial properties, lightweight, and cheap and simple processing techniques, have proved to be better replacements for the conventional packaging materials such as metals, ceramics, and paper. The inherent barrier properties (mechanical and thermal), biodegradability, self-healing, and self-cleaning of those composites increase the shelf life of the packaged food items. Polymer/clay has considerable performance in the packaging of processed foods like cheese, meats, confectionary, cereals, boil-in-bag foods, and even for fruit juices and carbonated drinks as shown in Figure 1.10 [35]. Some thin films like TiO_2-coated polypropylene showing strong antibacterial activity against *E. coli* can reduce the risk of microbial growth on the surface of fresh-cut vegetables (cut lettuce). Some of the polymer films containing antimicrobial silver nanoparticles also act as good packaging materials. In this regard, bio-nanocomposites also play a significant role with much improved mechanical strength, barrier properties, and heat resistance as compared to pure polymers or their conventional composites. The addition of cellulose nanocrystals to pure polylactic acid (PLA) and poly(vinyl alcohol) proved to be encouraging for food packaging applications [36]. In another case study by Youssef et al. [37], prepared paper sheets (from rice straw; an agricultural waste) coated with 5%–10% polystyrene nanocomposites using TiO_2 nanoparticles (with and without Ag nanoparticles) were used. Ag nanoparticles as antimicrobial agents exhibited a considerable inhibitory influence on different bacterial species (*Pseudomonas, Staphylococcus aureus, Candida,* and *Staphylococcus*) [38,39].

1.3.6 Energy

Materials with high dielectric constant, optimum piezoelectric properties are often searched for their importance in energy storage and harvesting application (Figure 1.11) [40]. Low dielectric constant polymers, when blended with dielectric/piezoelectric ceramics nanofillers, form a good combination with requisite properties. It was observed that 1D nanofillers could improve the energy density of the polymer nanocomposites by increasing the interfacial polarization. Zhang et al. [15] developed a sandwich-structured nanocomposite with $BaTiO_3$ nanofibers as fillers in the middle layer sandwiched between $BaTiO_3$ nanoparticles as fillers in the outer layers with a simultaneous increase in the dielectric and breakdown strength of the composites. Accordingly, the

FIGURE 1.10
Polymer nanocomposites used in different food packaging applications. (From Sorrentino, A. et al. *Trends Food Sci. Technol.*, 18, 84–95, 2007.)

energy density also increased. Similarly, PVDF-HFP/TiO$_2$ nanorod array nanocomposites exhibited a relatively high energy density of 17.1 Jcm^{-3} at 509 MV/m, even a much higher energy density of 31.2 Jcm^{-3} under a small loading of 3 vol% TiO$_2$ nanofibers in BaTiO$_3$ nanoparticles (fillers) incorporated in a PVDF-HFP matrix [41]. Flexible polymer-based nanocomposites suitable for energy harvesting serve as the new generation functional materials [42]. A stretchable nanogenerator constituting $(1-x)\{Pb(Mg_{1/3}Nb_{2/3})O_3\}-x\{PbTiO_3\}$ (PMN-PT) microparticles and MWCNTs within polydimethylsiloxane (PDMS) polymer matrix was found to generate excellent power output (4 V and 500 nA) and a remarkable stretchability of 200% [43].

1.3.7 Bio-Medical

Polymer nanocomposites form the basic building blocks of life systems starting right from bone (a combination of ceramic phosphate crystallites

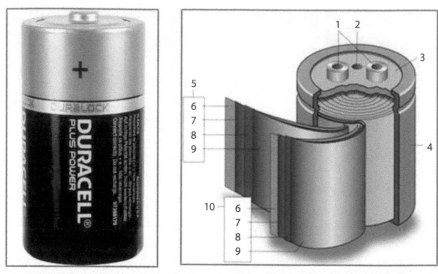

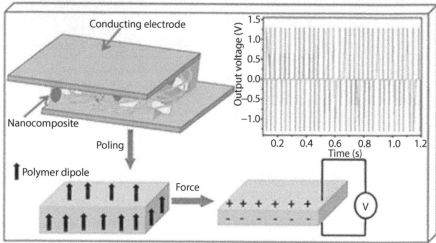

FIGURE 1.11
PNCs used in energy storage and harvesting applications. (From Ponnamma, D. and Al-Maadeed, M.A.A., *Sustain. Energ. Fuels.*, 3, 774–785, 2019.)

and collagen fibers forming strong and dense cortical bone or spongy shear resistant cancellous bone), teeth (enamel, cementum, and dentin containing different volume fractions of hydroxyapatite crystals along with collagenous or non-collagenous proteins), or wood (consisting of cellulose and lignin). Hence, polymers blended with other nanoparticles open up great avenues for a multitude of applications in a biomedical field such as tissue engineering, bone replacement/repair, dental applications, controlled drug delivery, and many more [44]. Some of the applications of polymer

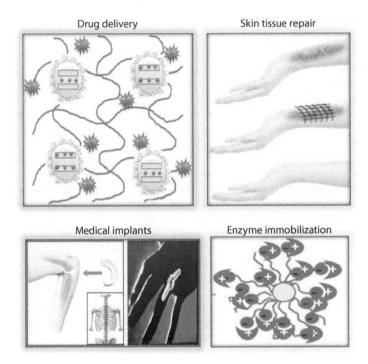

FIGURE 1.12

Polymer nanocomposites for different biomedical applications. (From Gârea, S.A. et al., Clay–polymer nanocomposites for controlled drug release, in *Clay-Polymer Nanocomposites*, , Elsevier, 2017, pp. 475–509.)

nanocomposites are shown in a representative Figure 1.12 [45]. A class of three-dimensionally hydrated polymeric networks, usually cross-linked with each other by nanostructures, popularly known as nanocomposite hydrogels, portray themselves as potential candidates that can mimic the physical, chemical, electrical, and biological characteristics of biological tissues. Hence, they encroach a significant portion in biomedical applications especially in regenerative medicine, drug delivery, biosensors, and bioactuators. CNTs and graphene-based nanocomposite hydrogels bestowed with high mechanical strength, electrical conductivity, and optical properties are best for use in tissue engineering scaffolds, conductive tapes, biosensors, and drug delivery systems [46]. On the same platform lies the biosensing devices which can detect small molecules such as glucose, cholesterol, lactate, and urea contents at very low concentrations (pico or nano-levels). Zou et al. in 2010 developed a magnetic biosensor with gold and magnetic nanoparticles as fillers in a polymer matrix which exhibited higher enzyme loading capability and enhanced biosensing capability than conventional counterparts [47]. Several magnetic polymer nanocomposites have been used in biomedical and environmental applications [48].

References

1. Camargo, P. H. C., K. G. Satyanarayana, and F. Wypych. Nanocomposites: Synthesis, structure, properties and new application opportunities. *Materials Research* 12.1 (2009): 1–39.
2. Hussain, F., M. Hojjati, M. Okamoto, and R. E. Gorga. Polymer-matrix nanocomposites, processing, manufacturing, and application: An overview. *Journal of Composite Materials* 40.17 (2006): 1511–1575.
3. https://ask.learncbse.in/t/how-is-the-following-resin-intermediate-prepared-and-which-polymer-is-formed-by-this-monomer-unit/10687
4. Hartono, A., M. Djamal, S. Satira, Herman, and Ramli. Preparation of PVDF film using deep coating method for biosensor transducer applied. 2013 3rd *International Conference on Instrumentation, Communications, Information Technology and Biomedical Engineering (ICICI-BME)*, pp. 408–411, 2013.
5. https://www.brainkart.com/article/Polymer-Structure_6365/
6. Dash, M., F. Chiellini, R. M. Ottenbrite, and E. Chiellini. Chitosan—A versatile semi-synthetic polymer in biomedical applications. *Progress in Polymer Science* 36.8 (2011): 981–1014.
7. Deshkulkarni, B., L. R. Viannie, S. V. Ganachari, N. R. Banapurmath, and A. Shettar 2018, June. Humidity sensing using polyaniline/polyvinyl alcohol nanocomposite blend. In *IOP Conference Series: Materials Science and Engineering* 376.1 (2018): 012063. IOP Publishing.
8. Sen, T., S. Mishra, and N. G. Shimpi. Synthesis and sensing applications of polyaniline nanocomposites: A review. *RSC Advances* 6.48 (2016): 42196–42222.
9. Guo, J., J. Long, D. Ding, Q. Wang, Y. Shan, A. Umar, X. Zhang, B. L. Weeks, S. Wei, and Z. Guo. Significantly enhanced mechanical and electrical properties of epoxy nanocomposites reinforced with low loading of polyaniline nanoparticles. *RSC Advances* 6.25 (2016): 21187–21192.
10. Wan, M., R. R. Yadav, D. Singh, M. S. Panday, and V. Rajendran. Temperature dependent ultrasonic and thermo-physical properties of polyaniline nanofibers reinforced epoxy composites. *Composites Part B: Engineering* 87 (2016): 40–46.
11. Reddy, K. R., K. P. Lee, Y. Lee, and A. I. Gopalan. Facile synthesis of conducting polymer–metal hybrid nanocomposite by in situ chemical oxidative polymerization with negatively charged metal nanoparticles. *Materials Letters* 62.12–13 (2008): 1815–1818.
12. Xu, X., Q. Wang, H. C. Choi, and Y. H. Kim. Encapsulation of iron nanoparticles with PVP nanofibrous membranes to maintain their catalytic activity. *Journal of Membrane Science* 348.1–2 (2010): 231–237.
13. Arbatti, M., X. Shan, and Z. Y. Cheng. Ceramic–polymer composites with high dielectric constant. *Advanced Materials* 19.10 (2007): 1369–1372.
14. Bai, Y., Z. Y. Cheng, V. Bharti, H. S. Xu, and Q. M. Zhang High-dielectric-constant ceramic-powder polymer composites. *Applied Physics Letters* 76.25 (2000): 3804–3806.
15. Zhang, X., Y. Shen, B. Xu, Q. Zhang, L. Gu, J. Jiang, J. Ma, Y. Lin, and C. W. Nan. Giant energy density and improved discharge efficiency of solution-processed polymer nanocomposites for dielectric energy storage. *Advanced Materials* 28.10 (2016): 2055–2061.

16. Abraham, R., S. P. Thomas, S. Kuryan, J. Isac, K. T. Varughese, and S. Thomas. Mechanical properties of ceramic-polymer nanocomposites. *Express Polymer Letters* 3.3 (2009): 177–189.
17. Harito, C., D. V. Bavykin, B. Yuliarto, H. K. Dipojono, and F. C. Walsh Polymer nanocomposites having a high filler content: Synthesis, structures, properties, and applications. *Nanoscale* 11.11 (2019): 4653–4682.
18. Karami, P., S. S. Khasraghi, M. Hashemi, S. Rabiei, and A. Shojaei. Polymer/nanodiamond composites-a comprehensive review from synthesis and fabrication to properties and applications. *Advances in Colloid and Interface Science* 269 (2019): 122–151.
19. Kumar, A., K. Sharma, and A. R. Dixit. A review of the mechanical and thermal properties of graphene and its hybrid polymer nanocomposites for structural applications. *Journal of Materials Science* 54.8 (2019): 5992–6026.
20. Gantayat, S., D. Rout, and S. K. Swain. Carbon nanomaterial–reinforced epoxy composites: A review. *Polymer-Plastics Technology and Engineering* 57.1 (2018): 1–16.
21. Moran, J. I., L. N. Ludueña, V. T. Phuong, P. Cinelli, A. Lazzeri, and V. A. Alvarez. Processing routes for the preparation of poly (lactic acid)/cellulose-nanowhisker nanocomposites for packaging applications. *Polymers and Polymer Composites* 24.5 (2016): 341–346.
22. Thomas, M. G., E. Abraham, P. Jyotishkumar, H. J. Maria, L. A. Pothen, and S. Thomas. Nanocelluloses from jute fibers and their nanocomposites with natural rubber: Preparation and characterization. *International Journal of Biological Macromolecules* 81 (2015): 768–777.
23. Salehpour, S., M. Jonoobi, M. Ahmadzadeh, V. Siracusa, F. Rafieian, and K. Oksman. Biodegradation and ecotoxicological impact of cellulose nanocomposites in municipal solid waste composting. *International Journal of Biological Macromolecules* 111 (2018): 264–270.
24. Voevodin, A.A., J. P. O'Neill, and J. S. Zabinski. Nanocomposite tribological coatings for aerospace applications. *Surface and Coatings Technology* 116 (1999): 36–45.
25. Joshi, M. and U. Chatterjee. Polymer nanocomposite: An advanced material for aerospace applications. In S. Rana, R. Fangueiro (Eds.), *Advanced Composite Materials for Aerospace Engineering* (pp. 241–264). Woodhead Publishing, Amsterdam, the Netherlands, 2016.
26. Nair, S., M. K. Pitchan, S. Bhowmik, and J. Epaarachchi. Development of high temperature electrical conductive polymeric nanocomposite films for aerospace applications. *Materials Research Express* 6.2 (2018): 026422.
27. Kausar, A., I. Rafique, and B. Muhammad. Aerospace application of polymer nanocomposite with carbon nanotube, graphite, graphene oxide, and nanoclay. *Polymer-Plastics Technology and Engineering* 56.13 (2018): 1438–1456.
28. Mrazova, M. Advanced composite materials of the future in aerospace industry. *Incas Bulletin* 5.3 (2013): 139.
29. http://www.ewriteright.in/tutorial/writers6.php.
30. Chirayil, C. J., J. Joy, H. J. Maria, I. Krupa, and S. Thomas. Polyolefins in automotive industry. In M. Al-Ali AlMa'adeed, I. Krupa (Eds.), *Polyolefin Compounds and Materials* (pp. 265–283). Springer, Cham, Switzerland, 2016.
31. Garces, J.M., D. J. Moll, J. Bicerano, R. Fibiger, and D. G. McLeod. Polymeric nanocomposites for automotive applications. *Advanced Materials* 12.23: 1835–1839.

32. Hollaway, L.C. Polymers, fibres, composites and the civil engineering environment: A personal experience. *Advances in Structural Engineering* 13.5 (2010): 927–960.
33. Moy, S. Advanced fiber-reinforced polymer (FRP) composites for civil engineering applications. In N. Uddin (Ed.), *Developments in Fiber-Reinforced Polymer (FRP) Composites for Civil Engineering*. Woodhead Publishing, Oxford, 2013: pp. 177–204.
34. Gauvin, F. and M. Robert. Durability study of vinylester/silicate nanocomposites for civil engineering applications. *Polymer Degradation and Stability* 121 (2015): 359–368.
35. Sorrentino, A., G. Gorrasi, and V. Vittoria. Potential perspectives of bionanocomposites for food packaging applications. *Trends in Food Science and Technology* 18.2 (2007): 84–95.
36. Rhim, J. W., H. M. Park, and C. S. Ha. Bio-nanocomposites for food packaging applications. *Progress in Polymer Science* 38.10–11 (2013): 1629–1652.
37. Youssef, A. M., S. Kamel, and M. A. El-Samahy. Morphological and antibacterial properties of modified paper by PS nanocomposites for packaging applications. *Carbohydrate Polymers* 98.1 (2013): 1166–1172.
38. Liu, Y., B. Zwingmann, and M. Schlaich. Carbon fiber reinforced polymer for cable structures—A review. *Polymers* 7.10 (2015): 2078–2099.
39. De Azeredo, H.M.. Nanocomposites for food packaging applications. *Food Research International* 42.9 (2009): 1240–1253.
40. Ponnamma, D. and M. A. A. Al-Maadeed. Influence of $BaTiO_3$/white graphene filler synergy on the energy harvesting performance of a piezoelectric polymer nanocomposite. *Sustainable Energy and Fuels* 3.3 (2019): 774–785.
41. Yang, K., X. Huang, Y. Huang, L. Xie, and P. Jiang. Fluoro-polymer@ $BaTiO_3$ hybrid nanoparticles prepared via RAFT polymerization: Toward ferroelectric polymer nanocomposites with high dielectric constant and low dielectric loss for energy storage application. *Chemistry of Materials* 25.11 (2013): 2327–2338.
42. Fan, F.R., W. Tang, and Z. L. Wang. Flexible nanogenerators for energy harvesting and self-powered electronics. *Advanced Materials* 28.22 (2016): 4283–4305.
43. Jeong, C. K., J. Lee, S. Han, J. Ryu, G. T. Hwang, D. Y. Park, J. H. Park et al., A hyper-stretchable elastic-composite energy harvester. *Advanced Materials* 27.18 (2015): 2866–2875.
44. Hule, R. A. and D. J. Pochan. Polymer nanocomposites for biomedical applications. *MRS Bulletin* 32.4 (2007): 354–358.
45. Gârea, S. A., A. I. Voicu, and H. Iovu. Clay–polymer nanocomposites for controlled drug release. In K. Jlassi, M. M. Chehimi, S. Thomas (Eds.), *Clay-Polymer Nanocomposites* (pp. 475–509). Elsevier, Amsterdam, the Netherlands, 2017.
46. Gaharwar, A. K., N. A. Peppas, and A. Khademhosseini. Nanocomposite hydrogels for biomedical applications. *Biotechnology and Bioengineering* 111.3 (2014): 441–453.
47. Zou, C.Y. Fu, Q. Xie, and S. Yao. High-performance glucose amperometric biosensor based on magnetic polymeric bionanocomposites. *Biosensors and Bioelectronics* 25.6 (2010): 1277–1282.
48. Kalia, S., S. Kango, A. Kumar, Y. Haldorai, B. Kumari, and R. Kumar. Magnetic polymer nanocomposites for environmental and biomedical applications. *Colloid and Polymer Science* 292.9 (2014): 2025–2052.

2

Fabrication and Characterization of Nanocomposites

2.1 Blending of Nanofillers

2.1.1 Introduction

Generally, in polymer nanocomposites (PNCs), nanofillers have been used to reduce/modify the miscibility and morphology between the composite constituents and subsequently enhance the diverse material properties, i.e., electrical, mechanical, barrier, flame retardancy, thermal, etc. As mentioned above, it primarily depends on the choice of nanofillers, the mode of their distribution into the matrix, and their interaction mechanism. On the basis of dimension, these nanofillers can roughly be categorized into one- (e.g., nanographene platelets, montmorillonite clay, layered silicates, etc.), two- (carbon nanotubes, cellulose whiskers, silver/gold nanotubes, etc.), and three- (quantum dots, carbon black, ZnO, Fe_2O_3, and TiO_2 nanocrystals/particles, etc.) dimensional nanoscale fillers (Figure 2.1). When these fillers reinforced into the polymer matrix, they produce some unique sublime properties elevating them as potential candidates for microelectronics, sensors, catalysts, biomedical, coatings, and energy applications. Because of the high surface activity of the nanofiller and the tendency to aggregate, often the particles create filler-clusters of micrometer size which is one of the major problems in preparing PNCs. In this section, some frequently used blending as well as fabrication methods are described.

2.1.2 Mechanical Mixing

2.1.2.1 Ball Milling

Mechanical ball milling, one of the environmentally and economically sustainable solid-state mixing methods, pretends that it can be used as an alternative technique to achieve homogeneous dispersion of nanofillers into polymer matrices. As this process works at low temperatures in the absence of a solvent and with any type of polymer matrix, it can outrank other techniques containing within itself several advantages, such as the

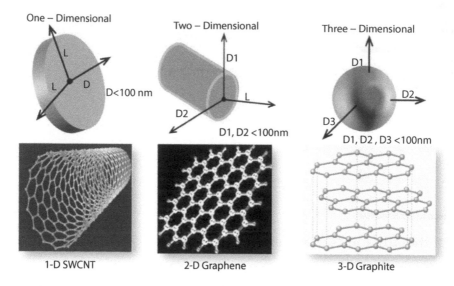

FIGURE 2.1
Classification of nanofillers on basis of dimension.

control over the degradation processes caused by high operational temperatures, compatibilization of immiscible blends, potent reduction of environmental disposal, the treatment of waste disposal and recycled materials, etc. Hence, ball milling in current years is well accepted as a novel green technique and attracts many researchers to explore the route for the preparation of advanced functional materials. It is basically following the top-down approach in which the filler/polymer powders are taken in a container (bowl) at proper ratio with the solid balls and is closed tight (Figure 2.2) [1]. The commonly available bowl and the balls are made up of stainless/ tempered steel, tungsten-carbide, zirconium oxide, agate, silicon nitride, and sintered corundum. The balls are then made to rotate around a predefined axis to apply a force on the material, and by the rotation of the supporting disc and autonomous turning of the bowl, the centrifugal forces act on the balls and the powders. As the supporting disc and the bowl are turning in the opposite directions, the centrifugal forces are synchronized alternately and opposite. Hence, the milling media and the charged powders roll alternatively on the inner wall of the bowl, and are then lifted and thrown off across the bowl at high speed (Figure 2.2). As a result, the powders are grounded to nanometer scale. This method was successfully applied in the fabrication of carbon nanofiller-based nanocomposites [2–4]. By tuning up the milling intensity, number of milling tools, the frequency and energy of the collisions, time, powder charge, temperature, atmosphere, etc., the properties of the final product can be tailored. The excess temperature and impurity during the process can be controlled by providing an inert gas atmosphere to the container and cryocooling arrangements.

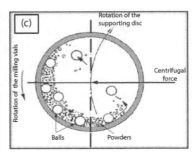

FIGURE 2.2
A snapshot of Fritsch (Pulverisette P5) (a) planetary ball mill, (b) bowl and balls, and (c) schematic picture of the ball motion during mixing. (From El-Eskandarany, M.S., The history and necessity of mechanical alloying, *Mechanical Alloying*, 2nd edn., Elsevier, pp. 13–47, 2015.)

2.1.2.2 Three Roll Mixing

Three roll milling or triple roll milling is a relatively simple, cheap, ease of handling, scalable, and high-volume milling technique widely utilized in industries to mix, refine, homogenize, or disperse a range of viscous materials like ceramics, carbon/graphite, pigments, plastisols, paints, etc. by making use of a shear force. This shear force results when three horizontally positioned rolls rotate in dissimilar speed and directions with each adjacent roller. A typical three roll mill and its operation in a schematic diagram is represented in Figure 2.3 [5]. These three rolls, i.e., feed, apron, and center roll rotate at higher speeds; however, the apron and feed roll rotate in the opposite direction to the center one. The materials to be dispersed are fed between the feed and center rolls and collected at the apron roll. The degree of dispersion

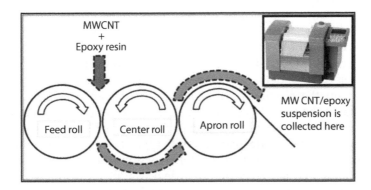

FIGURE 2.3
Schematic diagram of a typical three roll mill and its operation. The figure is reproduced from the reference. (From Olowojoba, G.B. and Fraunhofer, P., *Assessment of Dispersion Evolution of Carbon Nanotubes in Shear-Mixed Epoxy Suspensions by Interfacial Polarization Measurement*, Fraunhofer Verlag, Stuttgart, Germany, 2013.)

can be varied by controlling the shear intensity through suitably changing the roll speed and the spacing between them. This mixing method is often used to disperse carbon nanotubes (CNTs) into epoxy resin during the fabrication of CNT/epoxy nanocomposites to obtain enhanced electrical, thermal, and mechanical properties [6–9]. Raza et al. employed this technique for the dispersion of graphene in a silicone elastomer to improve the mechanical properties [10]. The influence of shear intensity on the property improvement was also investigated in CNT/epoxy-based nanocomposites [8,9].

2.1.3 Ultrasonic Mixing

Ultrasonic mixing has been proved to be one of the easiest, cost effective, and widely used process for the dispersion of nanoparticles in relatively low viscous fluids like water, ethanol, dimethylformamide (DMF), acetone, etc. using a sonicator. In both the sonicators by principle, the ultrasounds were generated through a series of attenuated waves via the medium, which assist in breaking the agglomerated nanoparticles into individual particles. Commonly, the bath sonicators are furnished with an ultrasound frequency of 20–23 kHz and operating power <100 W, but then the probe sonicators are equipped with a variable amplitude between 20% and 70% and a maximum power of 1500 W. Sonication is mostly used before the fabrication of the final PNCs' products. It is witnessed that many researchers use the sonication process particularly to break agglomerate CNTs into individual nanotubes prior to the preparation of CNT-based polymer nanocomposites [11,12]. Nevertheless, a few groups in recent years employed this technique to disperse CNTs directly into the epoxy matrix with slight modifications to prepare CNT/epoxy nanocomposites [11–14].

2.1.4 Magnetic Stirring

Magnetic stirring is one of the simplest techniques for the preparation of mixtures using a magnetic stirrer. This handy laboratory device employs a rotating magnetic field that leads to spinning the bar immersed in liquid quickly. Generally, the field is generated from a rotating magnet (or set of stationary electromagnets). The vortex effect from the magnet creates a collision between the particle-particle agglomeration and the bar, results in breaking down the electrostatic force responsible for particle agglomeration, and thus, produces partial dispersion of the nanoparticles into the polymer matrix. The stirring bars are available with typical sizes of a few mm to a few cm and coated with chemically inert Teflon or glass. Because of its small size, cleaning is much easier than the other devices without applying any lubricants, thus avoiding possible contamination. In recent times, almost all the magnetic stirrers are provided with heating arrangements for heating the liquid. However, the small size of the bars limits its use to small-scale experiments and low-viscous liquids. For large-scale experiments or thick suspensions, its use is usually coupled with other more potent mixing techniques.

During the preparation of an adhesive for wood application, Afolabi et al. have employed magnetic stirring for 2 h at 1250 rpm, followed by 90 minutes of sonication to assist good dispersion and interfacial interaction between carbon nanofillers and soy protein matrix [15].

2.1.5 Combination of Mechanical, Ultrasonic, and Magnetic Stirring

The above described mixing methods, i.e., ball milling, three roll milling, ultrasonication, and magnetic stirring are widely used to obtain the moderate dispersion/exfoliation of a variety of nanoparticles into suitable polymer matrices. However, reports claiming agglomeration/cluster-free dispersion utilizing any one of the above-listed methods individually are rarely available in literature. Nevertheless, a combination of each of those mixing methods is established as an effective way to ameliorate the degree of dispersion and subsequently enhance the properties. Agubra et al. [16] obtained satisfying clay exfoliation by combining the magnetic stirring/thinky mixing and three roll mixing to disperse montmorillonites into an epoxy matrix. Ghosh et al. [17] reported that the ultrasonic dual mixing breaks the strong multiwalled carbon nanotubes (MWCNTs) agglomerates and assists in obtaining a better and homogeneous distribution into the epoxy matrix. Moreover, ultrasonic dual mixing influences the thermal and tensile properties of MWCNT/epoxy nanocomposites. Similarly, Halder and group reported the influence of ultrasonic dual mixing on morphology, thermal, and mechanical properties of epoxy/SiO$_2$ or TiO$_2$ or ZnO$_2$ nanocomposites [18–21]. Chun et al. have studied the properties of conducting epoxy/graphene composites prepared by a high-speed mechanical stirrer and bath sonicator [22]. Lepico et al. adopted a solution blending method by combining sonication and magnetic stirring with a proper selection of solvents to disperse colloidal nanosilica in polymethylmethacrylate (PMMA) [23].

2.1.6 Melt Blending

Melt blending has become one of the popular techniques for the preparation of polymer nanocomposites, mostly adaptable thermoplastic and elastomeric polymer matrices. In this method, either the polymer is melted first and fillers of the desired amount are added to it or the polymer and filler are dry mixed, and then heated together in an inert atmosphere. Recent industrial processes, i.e., extrusion and injection molding can be categorized under this melt blending. As no organic solvents are used and being toxin-free, the technique is proved to be an eco-friendly, cost effective, and best for large-scale production. As a result, this technique has been adapted by many industries for mass production in the last few years. Employing the technique, Sheng et al. effectively reinforced treated Ti$_3$C$_2$ MXene nanosheets into a thermoplastic polyurethane matrix producing enhanced mechanical and thermal properties [24]. Su et al. also obtained encouraging mechanical properties in a polycarbonate-graft-graphene oxide composite prepared via melt blending [25]. Similarly, Sharika et al. employed melt blending to design a MWCNT-reinforced natural fiber/polypropylene nanocomposite with tunable electromagnetic interference (EMI) shielding performance [26]. However, the main drawback lies in the poor dispersion of nanofillers into the polymer matrix at higher loadings. Also because of strong shear forces, it may result in grapheme blocking in case of graphene-reinforced polymer composites, which deteriorate the conducting property of the composites [27,28].

2.1.7 Effect of Fictionalization and Grafting

The uniform dispersion of nanoparticles of suitable surface morphology within the polymer matrix is extremely demanding in the process of designing and fabrication of composites with novel macroscopic properties. In order to develop requisite physicochemical properties, control/modification of the surface morphology of the nanofillers is desirable utilizing their large surface to volume aspect ratio and strong surface effects. Hence, surface engineering of the nanofillers is essential for the deliberate manipulation of properties of PNCs. In this course of action of achieving homogeneous dispersion and subsequently enhanced interfacial interactions, the functionalization of nanofillers and grafting of polymers play a vital role. Generally, functionalization is of two types, i.e., chemical (covalent) and physical (non-covalent). A few of the possibilities of functionalization of CNTs are shown in Figure 2.4a [29]. When the chemical bond between the nanofiller and the matrix increases, the interfacial energy increases, resulting in the increase of toughness and strength. For example, the functionalization of CNTs leads to the adhering of hydroxyl and carboxyl function groups due to the breakage of the surface bonds. This may result in better adhering property and, subsequently, improve the bond strength between the matrix and CNT walls and, hence, the mechanical properties [30,31]. Bahamonde et al. prepared polysulfone nanocomposites with functionalized reduced graphene

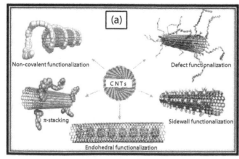
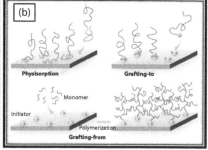

FIGURE 2.4
(a) A few possibilities of functionalization of CNTs. (From Hussain, C.M. and Mitra, S., *Anal. Bioanal. Chem.*, 399, 75–89, 2011.) (b) A schematic diagram of the grafting-from and grafting-to methods. (From Bhosale, R. et al. *J. Nat. Prod.*, 5, 124–139, 2015.)

oxide (rGO)/polysulfone (PSU) with improved dispersion and the interfacial interaction, which reflected in better mechanical and thermal properties as compared to unmodified rGO/PSU nanocomposites [32]. Moreover, grafting is another useful and versatile approach similar to functionalization to improve the polymer surfaces with the objective to overcome the issues like chemically inert surface, hydrophobicity, non-polar polymer surfaces, etc. which generally lead to adverse problems in adhesion, painting, coating, packaging, and so on. Again, grafting is majorly of two types; grafting-from and grafting-to. In the grafting-from method, the macromolecular backbone is chemically modified to introduce active sites that are capable of initiating functionality, whereas in the grafting-to approach, various bond linkages are used to develop for the surface modification.

Figure 2.4b shows a schematic diagram of the grafting-from and grafting-to methods [33]. For surface grafting, there are several methods, such as ultraviolet (UV) irradiation, plasma discharge, ozone, etc., that are successfully used by researchers. Recently, Pribyl et al. studied the synthesis, characterization, and challenges in making the polyethylene and nanosilica nanocomposites and the importance of grafting and functionalization [34]. Wei et al. have made PNCs with polymer-grafted nanoparticles which exhibit better crystallinity, ductility, Young's modulus, and tensile strength [35]. Wei et al. successfully prepared Styrene-Ethylene-Butadiene-Styrene copolymer-grafted graphene oxide/epoxy (Styrene-Ethylene-Butadiene-Styrene-g-GO/epoxy) nanocomposites through modification of the curing agents of a epoxy resin. They have reported that the flexural modulus, tensile strength, and impact strength of the nanocomposite (0.3 wt% Styrene-Ethylene-Butadiene-Styrene-g-GO) were 51.61%, 109.05%, and 139.32% higher than that of a virgin epoxy, respectively [36].

2.1.8 Solution Mixing

Solution mixing is one of the most common, easy, and efficient methods that is used for the synthesis/fabrication of polymer nanocomposites accompanying a wide range of polymers such as polyethylene, polyvinyl fluoride, polyvinyl alcohol, PMMA, epoxy, polyurethane, etc. Moreover, the technique requires no expensive equipment and operative protocols in spite of comprising a large number of, though quite simple, steps such as dispersion of fillers, incorporation of the polymer, and removal of the solvent. Initially, a solution is prepared by dissolving the selected polymer into a compatible solvent. By and large, the solvents like acetone, toluene, cyclohexane, tetrahydrofuran, dimethylformamide, and chloroform, etc. are seen to be used for the preparation of PNCs. Then the nanofillers with required wt% are added slowly into the solution through an hour-long constant stirring by sonication, magnetic stirring, reflux, or shear mixing. Finally, the solvent is removed by drying the solution, followed by molding to shape the composite as per the requirement for different measurements and characterizations. A schematic diagram is shown

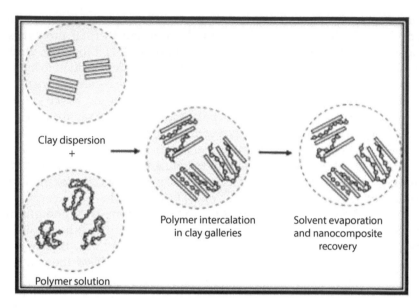

Clay dispersion +

Polymer solution

Polymer intercalation in clay galleries

Solvent evaporation and nanocomposite recovery

FIGURE 2.5
Synthesis of nanocomposites by dispersion in solution mixing. (From Mtibe, A. et al., Fabrication and characterization of various engineered nanomaterials, *Handbook of Nanomaterials for Industrial Applications*, Elsevier, pp. 151–171, 2018.)

in Figure 2.5 [37]. Numerous nanocomposites have been prepared through this method by many research groups worldwide effectively and efficiently to improvise the performance of the nanocomposites [38–40]. For example, Djahnit et al. obtained thermally stable a zinc oxide/polymethylmethacylate nanocomposite with UV shielding capability and optical transparency [41]. Gong et al. fabricated temperature-independent piezoresistive sensors utilizing CNT/epoxy resin nanocomposites prepared through the solution method [42]. However, the use of expensive solvents and their disposal create a hindrance to scale up and for their eventual adaptation by industries. Nevertheless, to address this issue, distilled water is used as a solvent to prepare some carbon nanofiller/rubber nanocomposites [43,44].

2.2 Fabrication

2.2.1 Introduction

Owing to the growing utility of polymer nanocomposites in various sectors, a lot of preparation methods are developed and employed to fabricate numerous PNCs to date. Nevertheless, the selection of a suitable and

compatible process is the key in order to meet specific design and manufacturing challenges for the end-user applications. PNC fabrication being a
complex process, as it integrated by the factors of design, manufacturing,
and economics, it requires concurrent thought over various parameters,
such as reinforcement and matrix types and their compatibility, component
geometry and structure, production scale or volume, tooling or equipment
requirements, environmental aspects, and process economy. In the following
sections, some of the fabrication techniques of PNCs are discussed.

2.2.2 Hand Lay-up Method

Hand lay-up (often called wet lay-up) is a primitive, common, simple, and
least expensive open mold method suitable for the fabrication of a wide variety of polymer composites expanding from small- as well as large-scale pro-
ductions using low cost equipment. In spite of the low production volume
per mold, it is feasible to yield substantive amounts utilizing multiple molds.
Figure 2.6 shows a schematic picture of the hand lay-up equipment [45].
In this method, first a gel coat is applied to the open mold using a spray
gun in order to incur a quality surface. After curing the coat sufficiently, the
reinforcing material (carbon, polymeric, or glass fibers) is manually placed
into the open mold. The laminating resin is applied using a paint roller or
by pouring and brushing. To ensure the proper wetting of the fillers and
to obtain a uniform resin distribution and thickness, hand rollers are used
to remove the air bubbles and enhance the resin-reinforcement interaction.
Often the process is repeated to meet the demanded thickness of the composites. The mold is then closed and left to be cured under standard atmospheric

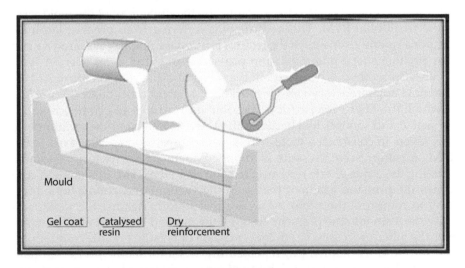

FIGURE 2.6
A schematic picture of the hand lay-up method. (From https://www.skyfilabs.com/project-ideas/
fabrication-of-glass-hybrid-fiber-epoxy-composite-material-using-hand-layup-mathod.)

conditions. The core materials particularly the low-density materials like foam, honeycomb, and end-grain balsa are generally used to stiffen the laminate. However, the laminating process of the matrix and the reinforcement is very crucial in the context of obtaining a quality composite product and needs expert skill. To date, numerous composites have been fabricated using the hand lay-up method. Utilizing this technique, Hallonet et al. recently prepared flax/epoxy composites suitable for the use of external strengthening of reinforced concrete (RC) structures [46]. Seretis et al. have investigated the post-curing effect on the thermal and mechanical behavior of graphene nanoparticle-reinforced hand lay-up glass fabric/epoxy nanocomposites [47]. They have observed that for a suitable combination of post-curing conditions, the nanocomposite laminates exhibited the optimal mechanical properties. Similarly, Budelmann et al. also reported the influence of process- (temperature, lay-up, and compaction speed) and materials-related (resin, age, draping/lay-up surface) parameters on material properties with regard to the tack properties of epoxy-impregnated carbon fibers [48].

2.2.3 Vacuum Resin Transfer Molding

The vacuum-assisted resin transfer molding (commonly known as VARTM) process has been established as one of the most reliable, low-cost, and hence, widely used methods for fabricating fiber-reinforced polymer (FRP) composites in a span of two decades since its development. Basically, it is a closed mold process, developed as a variant of the conventional RTM method, with a prime intention to overcome the design difficulties related to the large metal tools, and most importantly to reduce the cost. In terms of process advancement, a vacuum bag is introduced, replacing the upper half of the mold used in the traditional RTM. This technique is differing from prepreg laminate composite methods in which the resin is infused into the dry fabric on a mold near product shape under vacuum pressure and cured in an oven. A typical VARTM method is displayed in the schematic diagram in Figure 2.7 [49]. The mold appeared very similar to the hand lay-up open mold (Figure 2.6) in which the fiber preform/flow distribution media/peel ply/helical tubes are sealed and sandwiched between the VARTM mold and the vacuum bag. This favors to construct a mold with a larger dimension as compared to the RTM. A compressive pressure is supplied against fiber preform assembly utilizing the atmospheric pressure. In contrast to the RTM, this method consumes the pressure gradient (between the vacuum and atmospheric pressure) to compress the preform, secure the preform against the mold, and draw the resin into the preform. Besides, the method has a list of advantages over the other processes, i.e.: (i) low emission of volatile organic compounds, (ii) high product quality, (iii) repeatability, (iv) affordability, (v) clean handling of RTM process, (vi) scalability of open mold hand lay-up process, and (vii) flexible mold tooling design, and so on. However, the technique has some limitations such as flow distribution medium, vacuum bag, sealing

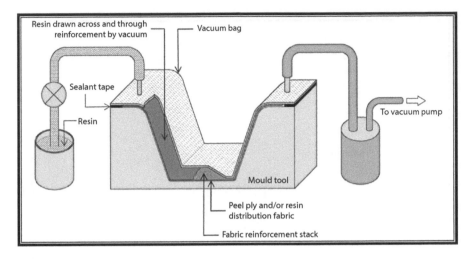

FIGURE 2.7
A typical VARTM method. (From https://www.nal.res.in/en/techniques/resin-transfer-moulding-processes.)

tape, peel ply, and resin tubing may not be reusable; high degree air leakage, limited resin injection pressure, and compression. Nevertheless, because of the number of advantages, the technique is widely used in many sectors, such as energy, marine, aerospace, in fracture building, and defense industries. For example, Kong et al. designed and manufactured an automobile hood for structural safety and stability using a flax/vinyl ester composite [50]. The group has also studied the nano ZnO-effect on the mechanical properties of ZnO/polyester-woven carbon-fiber composite fabrications using this method [51]. Nisha et al. have prepared electrostatic-bonded polyvinylidene fluoride (PVDF)-MWCNTs and glass fiber-reinforced polymer composites for structural health monitoring [52].

2.2.4 Filament Winding

Filament winding is one of the oldest manufacturing processes for the composites, used since the early 1960s and 1970s for the fabrication of fiber-reinforced pipes, streetlight poles, and pressure vessels. This technique has progressively turned out as a fast, cost-effective method to create lightweight, high-performance structures including driveshafts and golf clubs, oars/paddles, yacht masts, small aircraft fuselages, bicycle rims and forks, car wheels and pressure vessels, spacecraft structures, cryogenic fuel tanks for spacecraft, liquid propane gas tanks, and firefighter oxygen bottles owing to its suitability to automation and the escalated use of robotics and digital technologies. Over the years, several composite manufacturing industries worldwide such as ENERCON (Aurich, Germany), Cygnet Texkimp (Northwich, UK), MF Tech (Argentan, France), CIKONI (Stuttgart, Germany),

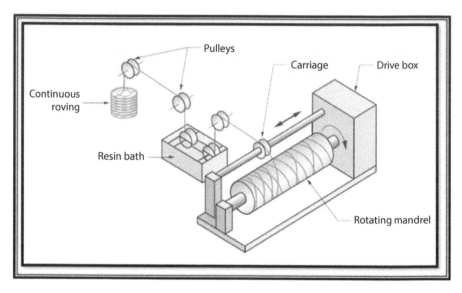

FIGURE 2.8
A typical filament winding machine. (From https://www.pinterest.com/pin/702631979334405930.)

and Murata Machinery Ltd. (Kyoto, Japan) are using filament winding very effectively and efficiently. Basically, this process impregnated the fibers in a resin bath prior to their application on a rotating mandrel (tool) while keeping them in tension. A typical filament winding machine is given in Figure 2.8 [53]. Besides wet winding, the processes have developed to make use of prepreg tapes and towpreg/dry fibers, while the latter is serving as preforms for liquid molding. The filament winding can also produce tailored-laminates to meet a range of loads efficiently by varying the angle of fiber/tape placement. In recently developed filament winding systems, the mandrel and fiber feed, both, can be moved and rotated through robots. These robots significantly cooperate to raise the accessibility range for fiber winding, enabling larger structures with more complex fiber-shapes and layups. Several simulation/theoretical models on filament winding are also described for thermoset/thermoplastic matrix composites, relating the processing conditions (i.e., processing speed, pressure, and temperature) to parameters of interest (i.e., viscosity, manufacturing stresses and strains, composite temperature, void sizes, etc.) [54]. Nevertheless, still there is ample scope to improve this technique to address some of the drawbacks such as: (i) the method is preferably used only to convex-shaped components, (ii) high cost of mandrel for large components, (iii) external surface of the components are cosmetically unattractive, and (iv) high viscosity resins create health and safety concerns. More recently, Weise et al. developed a pilot-scale melt-spinning process to fabricate graphene-modified multifilament yarns at 1800 m/min winding velocities (highest yield at 5% (w/w) loadings) for the development of anti-static and flame-retardant textiles [55]. Similarly, Zhao et al. have used this technique to design a shockless smart releasing device based on shape memory polymer composites [56].

2.2.5 Pultrusion

Pultrusion is a very fast, systematic, economic, and continuous manufacturing process for the products having constant cross section, e.g., structural shapes, rod stock, pipe, beams, channels, gratings, tubing, golf club shafts, decking, fishing rods, and cross arms. This continuous molding process takes the reinforcement of polymers to form high-strength structural parts of continuous length and makes fiber-reinforced polymer nanocomposites. Fundamentally, it comprises two steps: (i) forming process and (ii) heat treatments to produce parts of uniform cross-sections/lengths. The reinforcements, e.g., continuous strand glass fiber, basalt fiber roving, carbon fiber, mat, surfacing veil, etc. are generally pulled (by a pulling system) through the impregnation system (resin bath and performers: series of steel dies) (Figure 2.9) [57]. The finished part geometry is regulated using a cut-off saw. After forming the polymers, the curing and heating process is carried out inside the die and heating zone arrangements. The shape of the profile is dependent on the die, whereas the forces/heat treatments determine the performance of those finished parts. The performer is involved in rearranging the fibers of required shape and controlling the amount of resin used. The surface finish and structure can be improved by adding surface veils and strand mats prior to when it is passed through a heated-steel die. In order to limit the volatile emissions, the resin impregnation area can be enclosed. However, the main disadvantages of this technique lie in the high

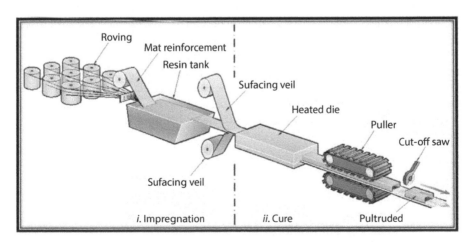

FIGURE 2.9
Pultrusion process. The figure has taken from the reference. (From Landesmann, A. et al., *Mater. Res.*, 18, 1372–1383, 2015.)

cost of heated die and the limitation of the process to fabricate constant or near constant cross-section components. In a review, Faiuz et al. have discussed the effects of aging after immersing kenaf-pultruded reinforced vinyl ester nanocomposites in solutions like distilled-water, seawater, and acidic solution on the morphology and the mechanical behavior [58]. Bowlby et al. developed some bio-composites by reinforcing carbon-based biochar particles in pultruded glass FRP composites with improved flexural strength [59]. Similarly, Chang et al. have studied the mechanical behavior of heat-treated pultruded-kenaf fiber-reinforced polyester composites [60]. Recently, Saenz-Dominguez's group has designed cellular composite structures out of die UV-cured pultrusion for automotive crash boxes [61].

2.3 Characterization of Nanocomposites

2.3.1 Introduction

Exploiting the uniqueness and superior quality of PNCs, several novel and high-performance composite materials have been developed by blending numerous nanofillers and polymers. In order to attain the required properties for specific application, it is very essential to characterize the PNCs through different characterization and property measurement techniques to elucidate the structure-property relationship. It enables a better understanding in various aspects of PNCs, such as the degree of filler dispersion in the polymer matrix and their alignment/orientation, filler-polymer interactions, effect of functionalization and grafting, morphology, and properties to ascertain their potential from an application view point. Hence, to generate a clear perception of PNCs, it is beneficial to use synergistic combinations of different characterization tools, i.e., thermogravimetric analysis (TGA), differential scanning calorimetry (DSC), X-ray diffraction (XRD), scanning electron microscopy (SEM), atomic force microscopy (AFM), transmission electron microscopy (TEM), X-ray photoelectron spectroscopy (XPS), infrared (IR)/Raman spectroscopy, nuclear magnetic resonance, electron spin resonance, etc. In this section, a few of these techniques are presented.

2.3.2 X-Ray Diffraction

X-ray diffraction is a powerful analytical and non-destructive technique primarily employed to extract ample information in connection with the atomic structure of a crystal, e.g., phase, crystal structure, inter-planner spacing, lattice parameters, composition, ordering, grain size, crystallinity, crystallite size, strain lattice defects, and dislocation substructure, etc. The incident monochromatic high energy X-ray beam generated from an electron gun is directed to interact with the atoms and get scattered at specific directions from the lattice planes. These elastically scattered X-rays undergo constructive interference, and the X-ray diffraction peaks are recorded as intensity versus 2θ. The inter-planner spacing (d-spacing) between two consecutive crystal planes is determined using Bragg's law; 2d sin θ = n λ, n is the order of reflection, d is the inter-planner spacing, λ is the X-ray wavelength, and θ is the scattering angle. In the research field of polymer nanocomposites, XRD is used to probe the dispersion of nanofillers (graphene, carbon nanotubes, nano-clays, and metal nano-oxides) in polymer matrices and also the interaction of these nanoparticles with the matrix. For example, the d-spacing between GO nano-sheets increased when amorphous carrageenan (Car) macromolecules were absorbed in GO nano-sheets through a hydrogen bond interaction, which is evidenced from the XRD study shown in Figure 2.10a [62]. Vaez et al. calculated the ratio of anatase and rutile phases of the crystalline TiO_2 in $N-TiO_2$/polyaniline (PANI) nanocomposites, which is considered to be an important parameter in the evolution of photocatalytic efficiency [Figure 2.10b]. It is also observed that the samples having a higher rutile phase exhibit higher photocatalytic activity [63]. The relation between the crystal structure/symmetry and mechanical property of a sample is experimentally demonstrated by many researchers. As an example, isotactic polypropylene showed enhanced elongation at break, higher toughness and impact strength in the presence of β-crystals, whereas in presence of α-crystals, it exhibits higher stiffness [64,65].

2.3.3 Electron Microscopy (SEM, TEM)

A variety of microscopic techniques, e.g., optical microscopy (~1 mm–1 μm), scanning electron microscopy (~100 μm–1 nm), transmission electron microscopy (10 μm–0.1 nm), and scanning tunneling microscopy (10 μm–0.1 nm) with different magnification/resolving power have been used to characterize the polymeric nanocomposites to extract direct and indirect evidences of

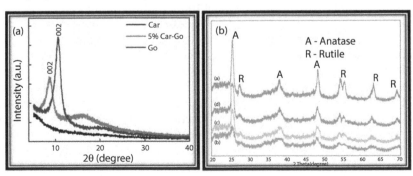

FIGURE 2.10
XRD patterns of (a) GO, Car, and 5% GO-Car nanocomposite films, (b) $N-TiO_2$/PANI nanocomposites at different $N-TiO_2$/PANI ratios. (From Zhu, W. et al., *Materials*, 10, 536, 2017; Vaez, M. et al. Polym. Compos., 39, 4605–4616, 2018.)

the structure of the nanofillers, filler-matrix adhesion, filler dispersion in a polymer matrix, impact of nanofillers on morphology and properties of the matrix, etc. Hence, the microscopic techniques play a crucial role in optimizing the nanocomposite properties in the process of developing nanocomposite science and technology. As an example, herein the use of two widely employed microscopic techniques (SEM, TEM) in polymer nanocomposite characterization has been provided very precisely.

SEM, a popular surface analytical tool is routinely used to estimate the filler distribution in a polymer matrix for PNCs. In this technique, a focused electron beam interacts with the specimen and scans the surface of the sample in a raster pattern. Generally, the SEM images are taken in two modes, i.e., secondary electron and backscattered electron. A secondary electron often reflects the topology of the sample surface, whereas the contrast in a backscattered electron is correlated with the element's atomic number. Roughly, the areas with a higher atomic number appear brighter, while the regions appear dark having elements of lower atomic number. Further, to estimate the chemical composition of the sample surface, energy dispersive X-ray (EDX) analysis can be done. It is beneficial to use SEM while investigating the filler dispersion and morphology of PNCs, as it warrants inspecting over a large sample surface as compared to TEM. In comparison to SEM, TEM provides relatively direct evaluation of PNC morphology with high precision images, dispersion state and adhesion, and the structure of the fillers. Even, the dispersion of an individual CNT in a polymer matrix can be noticed through TEM. Moreover, the morphology of PNCs comprising two- and three-dimensional fillers can be examined by TEM. Further, the information includes very local structural properties of the fillers and the filler-matrix interfaces. Besides, this technique has been intensively used to understand the morphology formation kinetics in many block copolymer-based hybrid nanocomposites. Often, this technique is use to extract information about the deformation mechanism, which is closely related to the mechanical performance of PNCs. Similarly, the agglomeration problem of nanofillers in PNCs apparently deteriorate the mechanical properties, and those agglomerations can be estimated nearly accurately from the TEM micrographs. Hence, the electron microscopic techniques play a vital role in providing a strong basis to elucidate the structure-property relationship in PNCs. A representative figure consisting SEM/TEM micrographs is given in Figure 2.11 [66,67]. The micrograph in Figure 2.11a shows the electrospun polyvinylpyrrolidone (PVP)/Au-nanoparticles (NPs) nanofibers are bead-free and uniform with an average fiber diameter of ~810 nm. The small shining spots in a high magnification SEM image [inset of Figure 2.11a] confirm the Au-NPs in the PVP nanofibers. The TEM image in Figure 2.11b displays spherical Au-NPs of an average diameter of ~5–20 nm which was dispersed throughout the PVP matrix. The elemental analyses performed by EDX [inset of Figure 2.11b] showed that C, N, O, and Au were the main elements of the PVP/Au-NP nanofibers [66]. Figure 2.11c–e present the

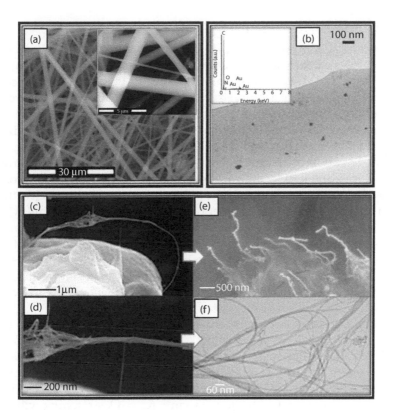

FIGURE 2.11
(a) SEM (inset: high magnification image) and (b) TEM (inset: EDX) micrographs of electrospun PVP/Au-NPs nanofibers; SEM micrographs of virgin [(c), (d)] and magnetically sensitive; (e) SWCNTs; and (f) TEM micrograph of virgin SWCNT. (From Deniz, A.E. et al., *Mater. Lett.*, 65, 2941–2943, 2011; Liu, M. et al., *Polymer*, 166, 81–87, 2019.)

SEM micrographs taken on the single-wall carbon nanotubes (SWCNT)/ polymer nanocomposites in the presence/absence of the magnetic field. Without the magnetic field, a bulk of the nanotubes are tangled between them due to the van der Waals forces that exist among different strands. Nevertheless, the magnetically sensitive nano-Fe_2O_3 particles are attached/ tethered to the ends/surfaces of the carbon nanotubes via the surfactants (the iron oxide bright particles identified by the arrows) [Figure 2.11e]. In a recent report, Zhang et al. observed bead-on-string morphology in PVDF/ BN nanofibers with average diameter increased from 245 nm to 328 nm at 0.5 wt% with the addition of modified-boron nitride nano-sheets (BNNSs). The EDX confirms the boron (B) and nitrogen (N) elements (the dark regions are the BNNSs materials). The morphology of the nanofiber composites provides indirect evidence of the good compatibility between the BNNSs (NH_2 functional groups) and the PVDF matrix [68].

2.3.4 Infrared and Raman Spectroscopy

The two vibrational spectroscopic techniques (infrared and Raman) are widely employed for the elucidation of information about different functional groups and their vibrational properties in PNCs. IR spectroscopy is more beneficial to study asymmetric vibrations of polar groups, whereas the Raman effect is an inelastic scattering mostly used for symmetric vibrations of non-polar groups. Both the vibrational modes/bands are characterized by their frequency, intensity, and band shape. The frequencies of these vibrational modes are mostly dependent on atomic masses, geometric arrangement of atoms, and their chemical bond strength. As IR radiation is promptly absorbed by functional groups of the polymer and the filler particles, the utilization of IR transmission spectroscopy is limited to ultra-thin films, whereas Raman spectroscopy can be applied to analyze thick polymeric composite samples. As a whole, the vibration spectral analysis is expected to furnish information of the interaction between the inorganic and organic phases, the state of intercalation/exfoliation in polymer/silicate nanocomposites, orientation of polymer chains and the anisometric nanofillers, the degree of filler dispersion and functionalization, etc.

Figure 2.12a,b show two representative fourier-transform infrared (FTIR) spectra of rGO/poly vinyl alcohol (PVA) film and functionalized MWCNT, respectively. The spectra of GO (Figure 2.12a) shows the characteristics bands at 3429 cm^{-1} (O–H stretching vibration of carboxyl groups and the absorbed water), 1385 cm^{-1} (skeletal vibrations of C–OH), 1270 cm^{-1} (skeletal vibrations of C–O–C), 1735 cm^{-1} (C = O stretching vibration), 1630 cm^{-1} (C = C skeletal stretching vibration), and 2800–3000 cm^{-1} (stretching vibration of C–H$_2$). With the addition of PVA, the stretching bands due to O–H and C = O become stronger and shifted to 3344 cm^{-1} and 1730 cm^{-1}, respectively, indicating the presence of H$_2$-bonding between the hydroxyl groups and the remaining oxygen-containing functional groups of PVA and rGO, respectively. The composite film also shows a CH/CH$_2$ deformation vibration band at 1300–1500 cm^{-1}. Overall, the results confirm the strong interfacial interaction between rGO and PVA [69]. Similarly, Yee et al. confirm the attachment of oxygen-containing functional groups on chemically functionalized MWCNTs from FTIR spectra of pristine and oxidized MWCNTs Figure 2.12b [70]. For example, the intensity of the O–H stretching vibration band of carboxyl groups at 3380 cm^{-1} increases after surface modification, indicating a high degree of covalent functionalization of MWCNTs by liquid phase oxidation using a H$_2$SO$_4$/HNO$_3$ mixture. Figure 2.12c [71] displays the Raman spectra of graphene/PANI composites. It is well known that graphene and graphene oxide have two characteristic bands, i.e., D-band at 1350 cm^{-1} due to sp^3 defects or edge area and G-band at 1598 cm^{-1} due to sp^2 hybridized carbon. Pure PANI exhibits bands at 1162 (C–H bending of the quinoid ring), 1338 (C–N^{+} stretching of the bipolaron structure), 1507 (N–H bending of the

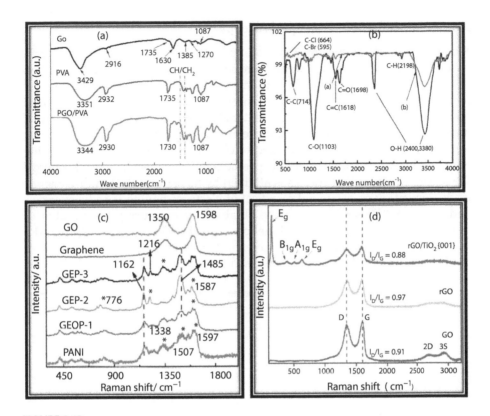

FIGURE 2.12
FTIR spectra of (a) GO, PVA, and rGO/PVA film, (b) pristine and functionalized MWCNT; Raman spectra of (c) graphene, GO, PANI, graphene/PANI composites, and (d) GO, rGO and rGO/TiO$_2$ {001} nanocomposite. (From Wang, T. et al. *RSC Advances*, 5, 88958–88964, 2015; Yee, M.J. et al. *Sci. Rep.*, 8, 17295, 2018; Wang, H. et al., *Nanoscale*, 2, 2164–2170, 2010; How, G.T.S. et al. *Sci. Rep.*, 4, 5044, 2014.)

bipolaronic structure), and 1597 cm^{-1} (C–C stretching of the benzenoid ring). The D/G ratio increases from 0.76 for GO to 0.83 for graphene, indicating the increased defects or edge areas by the reduction of GO. The results indicate that the reduction/dedoping–redoping processes affect the PANI chain structure greatly, which may influence the properties of the composite [71]. Similarly, How et al. [72] confirm the successful reduction of GO to rGO from the downshifting of G band from 1597 to 1587 cm^{-1}. The I$_D$/I$_G$ ratio increased from 0.91 to 0.97, further confirming the reduction of GO to rGO [Figure 2.12d] [72]. However, the I$_D$/I$_G$ ratio slightly decreases to 0.88 for the rGO/{001}TiO$_2$ nanocomposite due to the decrease in the sp^2 domain size of carbon atoms and the reduction of sp^3 to sp^2 carbon during the solvothermal process. Moreover, the E$_g$, B$_{1g}$, A$_{1g}$, and E$_g$ modes in the rGO/TiO$_2$ {001} nanocomposite reflect the presence of the TiO$_2$ anatase phase.

2.3.5 X-Ray Photoelectron Spectroscopy

X-ray photoelectron spectroscopy is one of the valuable modern-day techniques to characterize polymer nanocomposites particularly containing carbon fillers to identify/determine the structure, impurity elements on the surface, density of electronic states, chemical composition, and nature of the chemical bond. Basically, in this method, the material surface is irradiated by a beam of monochromatic X-ray photons, and the electrons emitted preferably from the top (~0–10 nm) surface are analyzed to obtain requisite information. As the electrons residing in any species in specific states have certain binding energy, the identification of that species is attained by measuring the binding energy. Each element gives rise to a distinct set of electrons having specific energies, and thus, an XPS spectrum can be obtained by measuring the number of these electrons as a function of kinetic energy. Herein, a representative XPS spectra taken from a report [73] is given in Figure 2.13. In this report, Naidek et al. have functionalized SWCNTs with the monomers 3-bromothiophene,

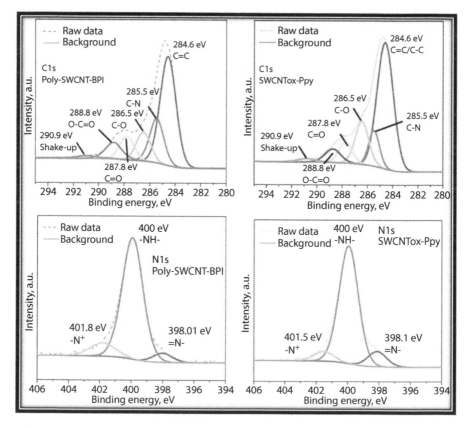

FIGURE 2.13
High-resolution C 1s and N 1s XPS spectra of Poly-SWCNT-BPI and SWCNT-Ox-Ppy. (From Naidek, N. et al. *New J. Chem.*, 43, 10482, 2019.)

3-acetylthiophene, and 1-(2-bromoethyl)-1H-pyrrole (BPI) by a one-step and room temperature reaction, and then synthesized covalent and non-covalent nanocomposites of SWCNT and polypyrrole (Ppy). The XPS spectra confirm the covalent attachment of the monomers on the surface of SWCNTs. Both the composites (covalent and non-covalent) exhibited similar chemical properties, and the Ppy is found to be in the conductive form in both composites. Wang et al. [69] prepared a flexible rGO/PVA film exhibiting superior microwave absorption behavior, and using XPS measurement, they confirmed the reduction process and the restoration of graphitic structure through the reduction process. Similarly, Huangfu et al. [74] and Wang et al. [75] have successfully employed the XPS technique to characterize Fe_3O_4/graphene-aerogel (thermally annealed)/epoxy EMI shielding nanocomposites and MWCNT-Fe_3O_4@ Ag/epoxy nanocomposites, respectively.

2.3.6 Brunauer-Emmett-Teller

Brunauer-Emmett-Teller (BET) is an ideal method for the measurement of the specific surface area of materials based on the theory that analyzes the physical adsorption of gas molecules on a solid surface. The BET theory is employed for the systems of multilayer adsorption to probe the gases that do not chemically react with material surfaces. N_2 is the most commonly used gaseous adsorbate for surface probing by BET methods. For an example, the BET analysis of PANI, PANI/charcoal (AC), and PANI/AC/Ni nanocomposites is provided in Figure 2.14 [76]. This figure shows a type-1 isotherm for PANI and type 4 isotherms for PANI/AC and PANI/AC/Ni nanocomposites. The BET surface area of PANI (21.9295 m^2g^{-1}) has doubled after adding charcoal, and increased further when Ni-nanoparticles are incorporated into the final nanocomposite. Furthermore, the structure of the materials synthesized contain mesopores, and their size decreases with increasing the constituents.

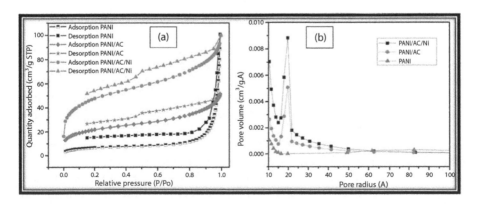

FIGURE 2.14
(a) BET surface area and (b) pore size distribution curves of PANI and PANI/AC and PANI/AC/Ni nanocomposites. (From Elanthamilan, E. et al., *Sustain. Energ. Fuels*, 2, 811–819, 2018.)

The pore volume also exhibited the same trend as the pore size [Figure 2.14b]. The increased surface area was responsible for its enhanced specific capacitance value [76]. Gouthaman et al. [77] have the surface area of the nanocomposite as 58.221 m^2/g from BET analysis and also confirmed the mesoporous surface of PNCs with the pore volume and pore radius determined to be 0.189 cm^3/g and 1.91 nm, respectively. Later they correlated a high surface area to efficiency of the nanocomposite dye removal capacity.

2.4 Mechanical Properties

2.4.1 Introduction

For decades, numerous studies have been carried out on the mechanical properties of PNCs, and it has been noticed that the incorporation of nanofillers into a polymer matrix firmly improved the mechanical properties under optimized processing conditions, i.e., a suitable processing method, different combinations/concentration of polymers and nanofillers, aspect ratio, etc. The mechanical testing of PNCs is vital to ascertain that the material complies with operational demands in accordance with industrial specifications, particularly to the automotive, aerospace, medical, defense, and consumer industries. Hence, the mechanical testing of PNCs necessitates the determination of mechanical factors i.e., tensile, impact, flexural, shear, compression, and fracture toughness.

2.4.2 Tensile Test

Tensile properties are basically understood as the force required to break a composite/plastic sample and the extent to which the specimen stretches/ elongates to its breaking point. These properties are determined through tensile testing, a destructive process usually described by an American Society for Testing and Materials (ASTM) standard (ASTM D 3039, ASTM D 638, and ASTM C 297). Specifically, ASTM D638 is recommended for randomly oriented, discontinuous, moldable, low reinforcement-volume composites. Instead, ASTM D3039 is applied for highly oriented and/or high tensile modulus FRP composites. Alignment is vital for composite testing applications, as the composites are anisotropic and brittle. Herein, the anisotropy means that the strength and properties of the material are highly dependent on the direction of the applied force/load. Thus, the tensile strength of a composite material is very high along the direction of the fiber orientation, while the tensile strength is much lower in any other direction. Specimens for tensile testing are usually dumbbell or dog bone-shaped or rectangular bar-shaped. Tensile tests provide information about tensile modulus, tensile

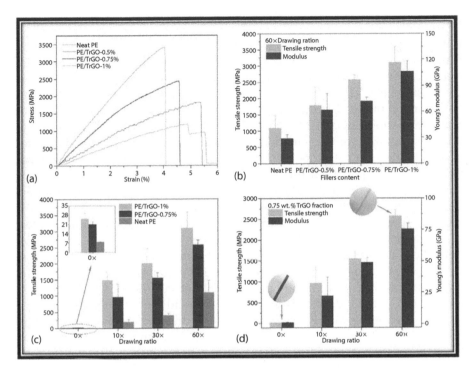

FIGURE 2.15
(a) Typical strain–stress curves, and (b) tensile strength and Young's modulus of pristine and
PE/TrGO composite films (60 × drawing ratio) with different filler fractions; (c) tensile strength
and (d) tensile strength and Young's modulus of composite films (0.75% filler fraction) at different drawing ratios. (From Pang, Y. et al. *ACS Nano*, 13(2), 1097–1106, 2019.)

strength both at yield and at break, tensile strain, elongation at yield, and
elongation at break in percent.

A representative figure displaying the tensile properties PE/TrGO composite films is shown in Figure 2.15 [78]. The strain–stress curves suggest
improvement in the Young's modulus and tensile strength, as the TrGO fraction increases because of higher intrinsic strength and stiffness of TrGO.
These improvements were majorly ascribed to the good dispersion of the
fillers and their large specific surface areas. Again, the Young's modulus and
tensile strength of pure PE and PE/TrGO composite films increase with a
higher drawing ratio [Figure 2.15c, d], which can be referred to the crystallinity of the drawn films [78]. Guo et al. have observed significant enhancement
in tensile strength (107.27 MPa) with low loading of (5.0 wt%) polyaniline
nanoparticles in epoxy nanocomposites [79]. Liu et al. observed a 9.8% and
9.7% increment in tensile strength in 1 wt% and 2 wt% magnetically aligned
SWCNT/polymer nanocomposites as compared to the values obtained for
the same composite without magnetic alignment [80]. Sandhya et al. also
reported improvement of tensile properties in phenol formaldehyde/rGO
composites with increasing rGO loadings [81].

2.4.3 Flexural Test

A flexural test mainly quantifies the force required to bend a beam under three-point loading conditions. This test is applicable to a wide range of materials, such as rigid and semi-rigid materials, resins, laminated fiber composites, etc. The commonly used flexural testing involves three-point and four-point bend testing according to International Organization for Standardization (ISO) 14,125, ISO 178, ASTM D 790, and ASTM D 6272 to ensure suitability for the intended application under various conditions for better insight into the properties of plastics, polymer composites, and large fiber-reinforced plates. In this test, the sample normally lies on a support span, the load is applied to the center by the loading nose, producing three-point bending at a fixed rate. The major parameters of this test are: (i) the support span, (ii) the speed of the loading, and (iii) the maximum deflection. For ASTM D790, the test is stopped before the specimen reaches 5% deflection, as it breaks afterwards. Flexural testing provides information on flexural stress at yield, flexural stress at break, flexural strain at yield, flexural strain at break, flexural modulus, stress/strain curves, and flexural stress at 3.5% (ISO) or 5.0% (ASTM) deflection. Alaaeddin et al. prepared PVDF/sugar palm fiber-reinforced nanocomposites with enhanced flexural strength (52 MPa) and flexural modulus (2151 MPa). This improvement is attributed to the fiber strength and interfacial bonding in the polyvinylidene fluoride (PVDF)-short sugar palm fiber (SSPF) composites [82]. Hung et al. fabricated multi-scaled-reinforced polymer composites by employing GO onto the surface of carbon fabrics through a novel electrophoretic deposition method. An improved flexural strength and modulus (by 20%–54% as compared to pristine composites) was obtained in GO/carbon-fiber reinforced polymer (CFRP) nanocomposites with only 0.5 wt% (at room temperature) and 0.25 wt% (at low temperature) of GO due to the effectiveness of crack deflection provided by GO on the surface of carbon fabrics, resulting in a high efficiency of energy dissipation by carbon fibre [83]. Singh et al. observed the enhancement of flexural strength in glass-FRP laminates using MWCNTs fabricated by hand lay-up and vacuum bagging technique [84].

2.4.4 Impact Test

An impact test ordinarily determines the response of the polymer, ceramic, and composite materials specimen to a suddenly applied stress. Explicitly, it is used to evaluate the brittleness, toughness, impact strength, and notch sensitivity of engineering materials to resist high-rate loading [85,86]. The ability to quantify the impact property is a great advantage in product liability and safety. The specimens to be tested include the notch configurations, such as keyhole notch, U-notch, and V-notch and the most common testing standards are ASTM D 695, ASTM D 3410, and ISO 14,126. The final stress-strain diagram provides information of yield point, yield strength, proportional limit, elastic limit, compressive strength, etc. The compression fixtures are designed/developed to meet the specific requirements of PNCs by furnishing precise alignment and guidance to prevent buckling. For example, Nor et al. added MWCNTs in bamboo/glass fiber hybrid composites to improve compression after impact and low velocity impact properties. With 0.5 wt% MWCNTs, the hybrid composites show low energy absorption, improved peak force, and deflection at a maximum of 9.21%, 36.23%, and 26.06%, respectively. Also, an increase of 23.67% was obtained on compression after impact strength [85]. Wen has observed a maximum value for impact strength of 14.5 kJ m^{-2} in SiO$_2$/grafted Poly(L-lactide) nanocomposites nearly 3.5 times higher than that of neat Poly(L-lactide) (4.2 kJ m^{-2}) [86].

2.4.5 Shear and Fatigue Test

Shear tests determine shear strain, shear stress, shear modulus, and failure mode. Since the awareness of the "deformable" mechanical properties of plastics and polymer composites is essential to extend their applications, shear testing can be used for quality control, comparative testing, and finite element analysis. A standard for the shear test is ASTM D3518. Other crucial mechanical tests, i.e., fatigue and fracture tests, accommodate the dynamic loads up to 2500 kN and are essential for composite materials, especially in demanding applications such as aerospace and wind power. In this test, the load frames provide high stiffness along with exceptional alignment that is needed for composite testing.

It has been observed by Park et al. that the in-plane shear modulus is higher in silica-mineralized nitrogen-doped CNT-reinforced poly(methyl methacrylate) nanocomposites as compared to the virgin PMMA and also CNT/PMMA nanocomposite [87]. Wang et al. reported excellent shear properties for graphene nanoplatelets/epoxy composites, particularly the shear strength of the composites with 0.75 wt% graphene exhibit a 102% increment compared to virgin epoxy adhesive [88]. The incorporation of 2 wt% of melamine functionalized-carbon nanotubes (M-CNTs) increased the inter-laminar shear strength of the carbon fiber/M-CNT/epoxy nanocomposite by 61%. Nevertheless, 2 wt% M-graphene nanoplatelets (GNP) increased the inter-laminar shear strength of the CF/M-GNP/epoxy nanocomposite by 219%. The higher reinforcing effect of the M-GNPs was attributed to differences in delamination resistance at the fiber-epoxy interfaces [89]. Bhasin et al. observed a greater enhancement in the fatigue crack growth resistance in the aligned GNPs/epoxy than the randomly oriented GNPs/epoxy nanocomposites. The enhancement is referred to various toughening mechanisms which retard the fatigue crack growth in the epoxy nanocomposites. Moreover, these mechanisms become more active when the GNPs are aligned normal to the direction of fatigue crack growth, resulting in higher fatigue resistance [90]. In a report by Bakis et al., it is seen that the fatigue crack propagation behavior mostly increases with the incorporation of different layered silicates, however, the highest fatigue crack propagation resistance was observed by shear stiff fluorohectorites with large (~10 times higher) lateral extension [91]. Similarly, Bourchak et al. studied the effect SWCNT (0.1 wt%) and graphene (0.1 wt%) nanoparticles have on the fatigue behavior of anti-symmetric glass fiber-reinforced polymer laminate. They have shown that the fatigue strength coefficient and the fatigue strength exponent of the nanocomposite increase up to 51% and 24%, respectively, with 0.1 wt% of SWCNTs loading, while the increment is 33% and 25%, respectively, in case of 0.1 wt.% of GNPs [92].

2.5 Electrical Properties

For many decades, lots of attention and efforts has been paid to develop high conducting polymer nanocomposites owing to their potential in many applications, such as anti-static coatings and films, electrically conducting adhesives, electromagnetic interference shielding thermal interface materials, and materials for electronic devices, etc. In order to achieve a high value of conductivity in those materials, different types of polymers, several conducting fillers with different concentrations, and many modified techniques have been adopted by a large number of researchers. Besides, using different nanofillers, particularly ceramic fillers, dielectric properties have been improved. Recently it was observed that the hybrid nanocomposites fabricated by employing multiple nanofillers showed further improvement of their electrical properties. For example, Aviles et al. studied electrical properties of vinyl ester nanocomposites filled with single nanofillers, i.e., CNTs, a few layers of a graphene shell of 3D cubic morphology, a few layers of graphene oxide platelets, and a hybrid combination of these nanofillers. They have observed that the electrical conductivity of the hybrid composites is higher than that of virgin polymer and the composites with a single filler [93]. Bagotia et al. also observed a similar effect of an optimum electrical conductivity of 1.91×10^{-1} S/cm in polycarbonate/ethylene methyl acrylate hybrid nanocomposites reinforced with multi-fillers of graphene: MWCNT in 1:3 ratio as shown in Figure 2.16a [94]. Nevertheless, Meeporn and Thongbai have obtained improved dielectric properties [Figure 2.16b] for PVDF nanocomposites enforced with hybrid fillers of Ag-nanoparticles and Ni ceramic particles [95]. Moreover, Bhawal et al. studied the effect of synthesis methods on the electrical conductivity of poly (ethylene methylacrylate)/carbon nanofibers composites [Figure 2.16c, d], and they have suggested that the melt blended nanocomposites exhibit better electrical conductivity with a lower percolation threshold [96]. Further, Xiang et al. studied the relation between phase transition and the electrical properties of high-density polyethylene/carbon nanotube and high-density polyethylene/graphene nanoplatelet composites of varying conductive network structures [97].

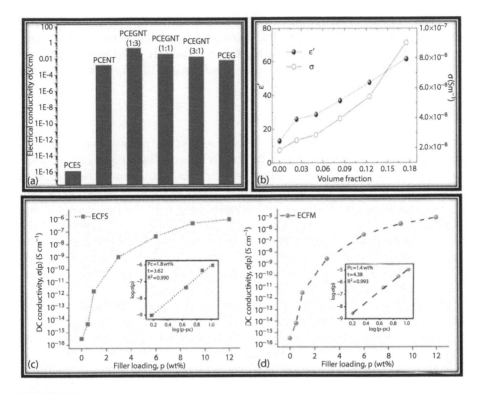

FIGURE 2.16
(a) Electrical conductivity of polycarbonate/ethylene methyl acrylate hybrid nanocomposites reinforced with multi-fillers of graphene: MWCNT in different ratios, (b) PVDF nanocomposites with hybrid fillers (Ag nanoparticles and nickelate ceramic), DC conductivity of carbon nanofibers (CNF)/ethylene methyl acrylate copolymer (EMA) hybrid composite of (c) solution mixing and (d) melt blending. (From Bagotia, N. et al., *Compos. Part B Eng.*, 159, 378–388, 2019; Meeporn, K. and Thongbai, P., *Appl. Surf. Sci.*, 481, 1160–1166, 2019 and Bhawal, P. et al. *Mater. Sci. Eng. B*, 245, 95–106, 2019.)

2.6 Thermal Properties

2.6.1 Introduction

The study of the thermal performance of the polymer nanocomposites is highly desirable to ensure the suitability of PNCs for various applications. Thermal behavior of the nanocomposites is witnessed to be better than both filler and virgin polymer, suggesting a synergetic improvement in the thermal properties. Nevertheless, the properties of the composite materials are primarily dependent on its two major components, i.e., nanofillers and matrix polymer. Hence, the evaluation of the thermal properties of the fillers, polymer, and the composites is very essential. The thermal properties, i.e., glass transition temperature (*Tg*), melting temperature, crystallization temperature, thermal conductivity, thermal expansion, decomposition temperature, etc. are generally measured by some widely used techniques like DSC, TGA, thermal conductivity measurements, etc. In this section, some of the techniques are presented briefly.

2.6.2 Thermal Gravimetric Analysis

TGA is useful in determining the decomposition temperatures of polymers and the organic content in PNCs. In this technique, the weight of the specimen is measured continuously either as a function of increasing temperature or time, using a high precision balance. The specimen is placed on a pan equipped with a microbalance. The sample along with the pan is heated at a controlled heating rate mostly in an inert atmosphere to get rid of the formation of side products during materials decomposition. The weight loss due to the degradation of polymer or any organic materials during the heating cycle is measured as a function of temperature. The following important information is extracted from the analysis of the TGA thermogram: (i) determination of the organophilization of the filler surface, (ii) ascertain the presence of any excess surface modification in the interlayer (using high resolution TGA), (iii) comparison of the thermal stability of different organically modified fillers, (iv) evaluation of thermal resistance of the fillers as function of time (dynamic TGA testing), and (v) analysis of physical adsorption or chemical reaction taking place (if any) on the surface of the organically modified fillers. Besides, it is often observed that thermal behavior of PNCs is environmentally sensitive, e.g., the thermal stability of the fillers seems to be better under a nitrogen atmosphere as compared to the air, which can be verified by recording the TGA curves at different temperatures.

The representative TGA thermograms in Figure 2.17 [98] clearly demonstrate the modification in the thermal stability of Ag/polymer

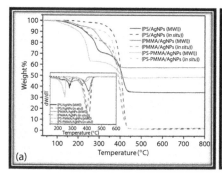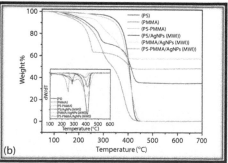

FIGURE 2.17
(a) TGA thermograms of polymer/AgNPs nanocomposites using microwave-irradiation (MWI) and *in situ* method and (b) TGA thermograms of neat polymers and polymer/AgNPs nanocomposites using MWI. (From Alsharaeh, E., *Materials*, 9, 458, 2016.)

nanocomposites with varying polymers [PMMA, polystyrene (PS), and PSMMA co-polymer] and synthesis methods (*in situ* bulk polymerization with and without microwave irradiation). As compared to the neat PSMMA, PS, and PMMA polymers, the nanocomposites, particularly PSMMA/Ag and PS/Ag composites prepared *in situ* technique, exhibited better thermal stability [98]. In another report on PANI/nanocrystalline cellulose/Ag-NPs nanocomposites prepared through *in situ* oxidative polymerization of aniline in the presence of nanocrystalline cellulose and Ag-NPs using sugar beet leaf extracts, Pashaei et al. studied the thermal stability with varying wt% of Ag-NPs. Besides quantifying the % of thermal degradation in three different stages (i.e., 15% wt loss at 1st stage ~40°C–141°C due to loss of unbound water, moisture, volatile impurities, 16% at 2nd stage ~268°C–527°C due to evaporation and degradation of nanocrystalline cellulose, structure of sugar beet leaf extract, and oxidation of the PANI structure, and 19% at 3rd stage ~516°C–781°C), and % char yield (~50%) is directly correlated with the potency of flame retardation, they have also studied the kinetics analysis of the thermal degradation [99]. Wang et al. have done a similar study on low-modified ZnO/vulcanized silicone rubber (RTV) nanocomposites to obtain the thermal decomposition temperature. From the thermal decomposition kinetics results, they have calculated the average activation energy as 144.29 and 146.78 kJ mol^{-1} for RTV and 1% ZnO/RTV, respectively [100].

2.6.3 Differential Scanning Calorimetry

Differential scanning calorimetry determines the temperature at which the local segmental motions of polymeric chains occur. The glass transition temperature (Tg), melting temperature (Tm), and crystallization temperature (Tc) can be obtained from this experiment. The commonly used DSC instruments contain a sample pan and a reference pan, and both pans are heated at the same rate and at the same time. The heat flow gradients between the pans are recorded and plotted as the function of temperature. A basic understanding of first-order and second-order transitions is essential in order to extract useful information from thermograms. Generally, the first-order transition involves latent heat (heat given off or absorbed) and heat capacity, e.g., melting is an endothermic process as polymers melt by absorbing heat, whereas crystallization is an exothermic process as polymers crystallize by giving off heat. In contrast, the second-order transition involves only the change in heat capacity. As polymers have high heat capacity at Tg, the small increase in temperature will increase the heat flow. Figure 2.18 [101] shows representative DSC curves for poly (lactic acid) nanocomposites filled with different fillers such as graphene oxide, silica nanoparticles, and silica-doped graphene oxide. Terzopoulou et al. extracted a large amount of useful thermodynamic information, e.g., glass transition temperature, melting temperature, heat capacity, rigid amorphous fraction, crystallinity, degree of crystallization, half crystallization time, etc. from the DSC thermogram [101].

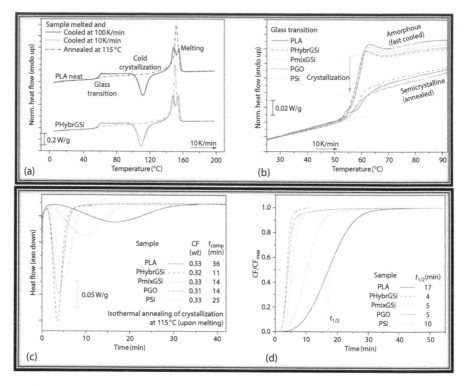

FIGURE 2.18
(a) DSC thermograms of PLA and PhybrGSi, (b) enlarged curves in the glass transition region, (c) time evolution of the relative crystallization fraction (CF/CF_{max}), and (d) half crystallization time. (From Terzopoulou, Z. et al. 2019, *Polymer*, 166, 1–12, 2019.)

Employing DSC, Viskadourakis et al. [102] have studied thermodynamical properties of poly actides (PLA)-based nanocomposite filaments, whereas Correia et al. [103] studied silk-elastin-like protein/CNT nanocomposites.

2.6.4 Thermal Conductivity

As mentioned in earlier sections, PNCs are universally used in a wide range of industrial applications owing to their lightweight, diverse functionality, low cost, and excellent chemical stability. However, the low thermal conductivity of generally used polymers limits their heat spreading capability and, hence, hinders their applications particularly in electronics fields. Thus, lots of efforts have been devoted to engineer PNCs with high thermal conductivity, heat exchangers and heat spreaders can be manufactured which can be applied in electronics, water, and energy industries. In order to improve the thermal conductivity, many researchers align the polymer chains by using mechanical stretching, electrospinning, and nanoscale templating. Nevertheless, blending polymers with conductive (thermal) fillers (metallic, ceramic, and carbonous) is considered to be the most common and easy technique used to enhance the thermal conductivity. Besides the polymer matrix and the fillers, the interaction among the fillers as well as between filler and matrix can also determine the thermal conductivity. Often the ceramic fillers (e.g., Al_2O_3, MgO, ZnO, AlN, BN, SiC) are preferred over the metallic ones (e.g., Cu- nanoparticles/nanowires, Al-fibers, Ag/Au nanoparticles), as there is a possibility that these metallic fillers may increase the electrical conductivity, which again prevents their applications as electrical insulators. Recently, nanostructured carbon fillers have been paid more attention as compared to the metallic and ceramic fillers on account of their high thermal conductivity (e.g., expanded graphite [EG] ~300 $W·m^{-1}K^{-1}$, graphene nanoplate ~1000–5000 $W·m^{-1}K^{-1}$, CNTs ~1000–3000 $W·m^{-1}K^{-1}$). Several methods, such as temperature wave method, hot plate and hot wire methods, differential scanning calorimetric method, and laser flash method, are used to measure the thermal conductivity of the composite samples. It is also noticed that several factors/parameters such as interfacial resistance and nanofiller distribution/dispersion/alignment affects the performance of thermal conductivity in PNCs. Wang et al. observed that the WCNT-Fe_3O_4@ Ag/epoxy nanocomposites exhibit adequate thermal conductivity with thermally conductive coefficient (λ) of 0.46 W/mK by introducing Fe_3O_4@ Ag nanoparticles into the polymer composite [75]. The thermal conductive value enhanced up to four times by reinforcing purified CNTs into the epoxy matrix [Figure 2.19a] [104]. Chen et al. fabricated PVDF nanocomposites containing orientated BNNSs [Figure 2.19b, c], which simultaneously exhibit high thermal conductivity enhancement (achieved ~16.3 $Wm^{-1}K^{-1}$ in the 18 μm thick nanocomposite film with 33 wt% BNNSs), excellent electrical insulation, and outstanding flexibility [105]. Moreover, Li et al. have studied the effects of defects (single-vacancy, double-vacancy, Stone-Wales, and multi-vacancy) on the interfacial thermal conductivity of graphene/ epoxy nanocomposites. Among them, the Stone-Wales and multi-vacancy defects significantly improve the interfacial thermal conductance, e.g.,

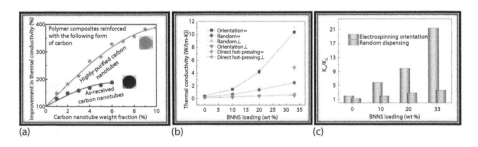

(a)   (b)   (c)

FIGURE 2.19

Thermal conductivity of (a) purified CNT/epoxy nanocomposite, and (b) and (c) PVDF/oriented BNNSs nanocomposites. (From Chen, J. et al., *Mater. Lett.*, 246, 20–23, 2019; Chen, J. et al., *ACS Nano*, 13, 337–345, 2018.)

when the Stone-Wales defect increased from 0% to 13%, interfacial thermal conductance increased from 135.54 to162.64 MW/m^2K [106].

2.6.5 Flame Retardancy

In day-to-day life, PNC materials provide numerous advantages to society. As most of the polymers are intrinsically combustible, an obvious demerit lies in the fact that PNCs are a fire hazard which includes ignitability, flame spread, flammability of volatile products, amount and rate of heat release, smoke obscuration, and toxicity etc. Hence, there is an urgent necessity to develop and improve the flame-retardant property in PNCs. To achieve it, flame retardants such as halogenated additives, endothermic additives, expandable graphite, phosphorus additives, and melamine derivatives are commonly utilized. It is often observed that to reach the required flame-retardant property, high loading of these additives is needed if a single flame retardant is used. This usually leads to the deterioration of the mechanical properties of the host polymer matrix. Afterwards, this drawback is addressed effectively by developing a synergetic flame-retardant system combining two or more additives, which minimizes the loading amount significantly. In this context, the combinations like antimony oxides/halogens and metal hydroxides/zinc borate are frequently used in various polymers. Recently, many research groups worldwide started using nanoscale additives to make polymer nanocomposites and expect further enhancement of the flame-retardant performance. Generally, the test methods such as TGA, the UL 94 vertical flame test, limiting oxygen index, cone calorimeter, and Steiner tunnel test are used to measure the flame-retardant performance.

Fang et al. have prepared epoxy nanocomposites by dispersing functionalized graphene oxide into the epoxy resin matrix. The graphene oxide was functionalized by introducing an organic component [piperazine and phytic acid] onto the surface of graphene oxide. As compared to the pure epoxy, the flame resistance of an epoxy nanocomposite is significantly improved, exhibiting a 42% decrease in peak heat release rate and 22% reduction in total heat release [107]. Lopez et al. prepared polyetherimide membranes filled with different wt% (1–5 wt.%) of GO using a solution casting method. They have observed that the flame-retardant properties were improved as the graphene oxide content was added into the polyetherimide membrane [108]. Les et al. reported about the intumescent flame-retardant polyketone nanocomposites containing multilayered expandable graphites incorporated with MWCNTs. Reduction of heat release rate, total heat release, and smoke production rate was observed when the intumescent expandable graphites were introduced in the nanocomposites. A synergistic improvement of thermal stabilities, i.e., high flame retardancy and self-extinguishing properties, are found by the addition of a small amount of CNTs in the composites [109]. Medina et al. studied the fire-retardant properties of recyclable nanocomposite foams of PVA, clay, and cellulose nanofibrils [110].

2.7 Biodegradability

Biodegradable and high-performance polymer and polymer nanocomposites are gradually emerging as the next generation sustainable materials for the future owing to their potential applications particularly in food packaging (e.g., clear films, trays, blister packs, compost sacks, and bags, etc.), agriculture (e.g., compostable materials in plant pots, mulching film, and yarn), medical sector (e.g., drug delivery, disposal gloves), and automobile sectors. In recent days, a great deal of interest and efforts have been given toward the design and development of biodegradable nanocomposites due to the increasing environment awareness/regulations and societal/economic concerns. The environmental parameters include water, temperature, soil, pH, moisture, sewage, compost, sludge, aeration, etc. The biodegradable polymers are roughly classified as bio-polymers [e.g., poly(hydroxyalkanote)], bio-derived polymers [from renewable sources: PLAs, starch plastics, cellulose esters; from petroleum products: aliphatic-aromatic polyesters; from mixed sources: bio-based epoxy or bio-based polyurethanes], and biodegradable polymers [e.g., polycaprolactone, PVA]. In this context, numerous PNCs are prepared expecting better performance of the biodegradable polymers by reinforcing/dispersing different nanoparticles/nanofillers. Considering the expected adverse environmental conditions during their applications as well as disposals at the end of their life cycle, it is essential to perform the biodegradation assessment of PNCs. The commonly used lab-based/on-field biodegradation test methods are: (i) analytical techniques (morphological, microscopic, gavimetric, spectroscopic, chromatographic, CO_2 and oxygen demand, physical, and chemical), microbiological techniques (direct cell count, pour plate, turbidity, etc.), enzymatic, and molecular techniques. Generally, the PNCs are investigated by measuring weight loss, molecular weight distribution, carbon dioxide production, tensile properties, biochemical oxygen demand, extent of fragmentation, enzyme assays, and eco-toxicity before and after degradation, and then are compared with the standards. Li et al. reported biodegradable ultra-long copper sulfide nanowire-reinforced poly(citrates-siloxane) nanocomposites elastomers, which showed broad-spectrum anti-bacterial activity against gram-positive/gram-negative bacterium *in vitro/in vivo*. Poly(citrates-siloxane) also exhibited tailored photoluminescent property and strong near-IR photothermal capacity which enabled the high-resolution *in vivo* thermal imaging and biodegradation tracking [111]. Similarly, Katerinopoulou et al. [112] have assessed the biodegradability of maize corn starch/glycerol glycerol/Na-montmorillonite and maize corn starch/glycerol/polyvinyl alcohol/montmorillonite hybrid nanocomposite films. Recently, Luzi et al. have reported a review article focusing on various polymer nanocomposites composed of natural fibers extracted from natural sources and biodegradable synthetic and/or natural polymers and their biomedical applications [113].

References

1. El-Eskandarany, M. S. The history and necessity of mechanical alloying. In M. S. El-Eskandarany (Ed.), *Mechanical Alloying*, 2n edn., Elsevier, Amsterdam, the Netherlands, 2015: pp. 13–47.
2. Witt, N., Y. Tang, L. Ye, and L. Fang. Silicone rubber nanocomposites containing a small amount of hybrid fillers with enhanced electrical sensitivity. *Materials & Design* 45 (2013): 548–554.
3. Alimardani, M., F. Abbassi-Sourki, and G. R. Bakhshandeh. An investigation on the dispersibility of carbon nanotube in the latex nanocomposites using rheological properties. *Composites Part B: Engineering* 56 (2014): 149–156.
4. Potts, J. R., O. Shankar, L. Du, and R. S. Ruoff. Processing–morphology–property relationships and composite theory analysis of reduced graphene oxide/natural rubber nanocomposites. *Macromolecules* 45.15 (2012): 6045–6055.
5. Olowojoba, G. B. and P. Fraunhofer. *Assessment of Dispersion Evolution of Carbon Nanotubes in Shear-Mixed Epoxy Suspensions by Interfacial Polarization Measurement*. Fraunhofer Verlag, Stuttgart, Germany, 2013.
6. Zhou, Y. X., P. X. Wu, Z. Y. Cheng, J. Ingram, and S. Jeelani. Improvement in electrical, thermal and mechanical properties of epoxy by filling carbon nanotube. *Express Polymer Letters* 2.1 (2008): 40–48.
7. Kim, J. A., D. G. Seong, T. J. Kang, and J. R. Youn. Effects of surface modification on rheological and mechanical properties of CNT/epoxy composites. *Carbon* 44.10 (2006): 1898–1905.
8. Thostenson, E. T. and T. W. Chou. Processing-structure-multi-functional property relationship in carbon nanotube/epoxy composites. *Carbon* 44.14 (2006): 3022–3029.
9. Rosca, I. D. and S. V. Hoa. Highly conductive multiwall carbon nanotube and epoxy composites produced by three-roll milling. Carbon 47.8 (2009): 1958–1968.
10. Raza, M. A., A. V. K. Westwood, A. P. Brown, and C. Stirling. Texture, transport and mechanical properties of graphite nanoplatelet/silicone composites produced by three roll mill. *Composites Science and Technology* 72.3 (2012): 467–475.
11. Ma, P. C., N. A. Siddiqui, G. Marom, and J. K. Kim. Dispersion and functionalization of carbon nanotubes for polymer-based nanocomposites: A review. *Composites Part A: Applied Science and Manufacturing* 41.10 (2010): 1345–1367.
12. Ajayan, P. M., L. S. Schadler, C. Giannaris, and A. Rubio. Single-walled carbon nanotube–polymer composites: Strength and weakness. *Advanced Materials* 12.10 (2000): 750–753.
13. Li, Q., M. Zaiser, and V. Koutsos. Carbon nanotube/epoxy resin composites using a block copolymer as a dispersing agent. *Physica Status Solidi (a)* 201.13 (2004): R89–R91.
14. Lau, K. T., M. Lu, C. K. Lam, H. Y. Cheung, F. L. Sheng, and H. L. Li. Thermal and mechanical properties of single-walled carbon nanotube bundle-reinforced epoxy nanocomposites: The role of solvent for nanotube dispersion. *Composites Science and Technology* 65.5 (2005): 719–725.
15. Afolabi, A. S., O. O. Sadare, and M. O. Daramola. Effect of dispersion method and CNT loading on the quality and performance of nanocomposite soy protein/CNTs adhesive for wood application. *Advances in Natural Sciences: Nanoscience and Nanotechnology* 7.3 (2016): 035005.

16. Agubra, V., P. Owuor, and M. Hosur. Influence of nanoclay dispersion methods on the mechanical behavior of E-glass/epoxy nanocomposites. *Nanomaterials* 3.3(2013): 550–563.
17. Ghosh, P. K., K. Kumar, and N. Chaudhary. Influence of ultrasonic dual mixing on thermal and tensile properties of MWCNTs-epoxy composite. *Composites Part B: Engineering* 77 (2015): 139–144.
18. Halder, S., P. K. Ghosh, M. S. Goyat, and S. Ray. Ultrasonic dual mode mixing and its effect on tensile properties of SiO_2-epoxy nanocomposite. *Journal of Adhesion Science and Technology* 27.2 (2013): 111–124.
19. Sinha, A., N. Islam Khan, S. Das, J. Zhang, and S. Halder. Effect of reactive and non-reactive diluents on thermal and mechanical properties of epoxy resin. *High Performance Polymers*, 30(10), pp. 1159–1168.
20. Halder, S., P. K. Ghosh, and M. S. Goyat. Influence of ultrasonic dual mode mixing on morphology and mechanical properties of ZrO_2-epoxy nanocomposite. *High Performance Polymers* 24.4 (2012): 331–341.
21. Ghosh, P. K., A. Pathak, M. S. Goyat, and S. Halder. Influence of nanoparticle weight fraction on morphology and thermal properties of epoxy/TiO_2 nanocomposite. *Journal of Reinforced Plastics and Composites* 31.17 (2012): 1180–1188.
22. Chun, W. W., T. P. Leng, A. F. Osman, and Y. C. Keat. The properties of epoxy/graphene conductive materials using high speed mechanical stirrer and bath sonicator. In Z. A. Ahmad, M. Y. Meor Sulaiman, M. A. Yarmo, F. A. Aziz, K. N. Ismail, N. S. Abdullah, Y. Abdullah, N. A. Rejab, M. Ahmadipour (Eds.), *Materials Science Forum*. Trans Tech Publications, Malaysia, 2017: Vol. 888, pp. 222–227.
23. Lepcio, P., F. Ondreas, K. Zarybnicka, M. Zboncak, O. Caha, and J. Jancar. Bulk polymer nanocomposites with preparation protocol governed nanostructure: The origin and properties of aggregates and polymer bound clusters. *Soft Matter* 14.11 (2018): 2094–2103.
24. Sheng, X., Y. Zhao, L. Zhang, and X. Lu. Properties of two-dimensional Ti_3C_2 MXene/thermoplastic polyurethane nanocomposites with effective reinforcement via melt blending. *Composites Science and Technology* 181 (2019): 107710.
25. Su, Y., G. Luan, H. Shen, B. Liu, S. Ran, Z. Fang, and Z. Guo. Encouraging mechanical reinforcement in polycarbonate nanocomposite films via incorporation of melt blending-prepared polycarbonate-graft-graphene oxide. *Applied Physics A* 125.6 (2019): 426.
26. Sharika, T., J. Abraham, S. C. George, N. Kalarikkal, and S. Thomas. Excellent electromagnetic shield derived from MWCNT reinforced NR/PP blend nanocomposites with tailored microstructural properties. *Composites Part B: Engineering* 173 (2019): 106798.
27. Singh, V., D. Joung, L. Zhai, S. Das, S. I. Khondaker, and S. Seal. Graphene based materials: Past, present and future. *Progress in Materials Science* 56.8 (2011): 1178–1271.
28. Du, J. and H. M. Cheng. The fabrication, properties, and uses of graphene/polymer composites. *Macromolecular Chemistry and Physics* 213.10–11 (2012): 1060–1077.
29. Hussain, C. M. and S. Mitra. Micropreconcentration units based on carbon nanotubes (CNT). *Analytical and Bioanalytical Chemistry* 399.1 (2011): 75–89.
30. Cha, J., J. Kim, S. Ryu, and S. H. Hong. Comparison to mechanical properties of epoxy nanocomposites reinforced by functionalized carbon nanotubes and graphene nanoplatelets. *Composites Part B: Engineering* 162 (2019): 283–288.

31. Maghsoudlou, M. A., R. B. Isfahani, S. Saber-Samandari, and M. Sadighi. Effect of interphase, curvature and agglomeration of SWCNTs on mechanical properties of polymer-based nanocomposites: Experimental and numerical investigations. *Composites Part B: Engineering* 175 (2019): 107119.
32. Peña-Bahamonde, J., V. San-Miguel, J. Baselga, J. P. Fernández-Blázquez, G. Gedler, R. Ozisik, and J. C. Cabanelas. Effect of polysulfone brush functionalization on thermo-mechanical properties of melt extruded graphene/polysulfone nanocomposites. *Carbon* 151 (2019): 84–93.
33. Bhosale, R. R., H. V. Gangadharappa, A. Moin, D. V. Gowda, and A. M. Osmani Grafting technique with special emphasis on natural gums: Applications and perspectives in drug delivery. *The Natural Products Journal* 5.2 (2015): 124–139.
34. Pribyl, J., B. Benicewicz, M. Bell, K. Wagener, X. Ning, L. Schadler, A. Jimenez, and S. Kumar. Polyethylene grafted silica nanoparticles prepared via surface-initiated ROMP. *ACS Macro Letters* 8.3 (2019): 228–232.
35. Wei, T., K. Jin, and J. M. Torkelson. Isolating the effect of polymer-grafted nanoparticle interactions with matrix polymer from dispersion on composite property enhancement: The example of polypropylene/halloysite nanocomposites. *Polymer* 176 (2019): 38–50.
36. Wei, L., X. Chen, K. Hong, Z. Yuan, L. Wang, H. Wang, Z. Qiao, X. Wang, Z. Li, and Z. Wang. Enhancement in mechanical properties of epoxy nanocomposites by Styrene-ethylene-butadiene-styrene grafted graphene oxide. *Composite Interfaces* 26.2 (2019): 141–156.
37. Mtibe, A., T. H. Mokhothu, M. J. John, T. C. Mokhena, and M. J Mochane. Fabrication and characterization of various engineered nanomaterials. In C. M. Hussain (Ed.), *Handbook of Nanomaterials for Industrial Applications.* Elsevier, Amsterdam, the Netherlands, 2018, pp. 151–171.
38. Silva, M., N. M. Alves, and M. C. Paiva. Graphene-polymer nanocomposites for biomedical applications. *Polymers for Advanced Technologies* 29.2 (2018): 687–700.
39. Chen, Z., X. J. Dai, K. Magniez, P. R. Lamb, D. R. de Celis Leal, B. L. Fox, and X. Wang. Improving the mechanical properties of epoxy using multiwalled carbon nanotubes functionalized by a novel plasma treatment. *Composites Part A: Applied Science and Manufacturing* 45 (2013): 145–152.
40. Guo, P., X. Chen, X. Gao, H. Song, and H. Shen. Fabrication and mechanical properties of well-dispersed multiwalled carbon nanotubes/epoxy composites. *Composites Science and Technology* 67.15–16 (2007): 3331–3337.
41. Djahnit, L., N. Sened, K. El-Miloudi, M. A. Lopez-Manchado, and N. Haddaoui. Structural characterization and thermal degradation of poly (methylmethacrylate)/zinc oxide nanocomposites. *Journal of Macromolecular Science, Part A* 56.3 (2019): 189–196.
42. Gong, S., D. Wu, Y. Li, M. Jin, T. Xiao, Y. Wang, Z. Xiao, Z. Zhu, and Z. Li. Temperature-independent piezoresistive sensors based on carbon nanotube/polymer nanocomposite. *Carbon* 137 (2018): 188–195.
43. Huang, N. J., J. Zang, G. D. Zhang, L. Z. Guan, S. N. Li, L. Zhao, and L. C. Tang. Efficient interfacial interaction for improving mechanical properties of polydimethylsiloxane nanocomposites filled with low content of graphene oxide nanoribbons. *RSC Advances* 7.36 (2017): 22045–22053.

44. Qiu, Y., A. Zhang, and L. Wang. Carbon black–filled styrene butadiene rubber masterbatch based on simple mixing of latex and carbon black suspension: Preparation and mechanical properties. *Journal of Macromolecular Science, Part B* 54.12. (2005): 1541–1553.
45. https://www.skyfilabs.com/project-ideas/fabrication-of-glass-hybrid-fiber-epoxy-composite-material-using-hand-layup-mathod.
46. Hallonet, A., E. Ferrier, L. Michel, and B. Benmokrane. Durability and tensile characterization of wet lay-up flax/epoxy composites used for external strengthening of RC structures. *Construction and Building Materials* 205 (2019): 679–698.
47. Seretis, G. V., S. F. Nitodas, P. D. Mimigianni, G. N. Kouzilos, D. E. Manolakos, and C. G. Provatidis. On the post-curing of graphene nanoplatelets reinforced hand lay-up glass fabric/epoxy nanocomposites. *Composites Part B: Engineering* 140 (2018): 133–138.
48. Budelmann, D., H. Detampel, C. Schmidt, and D. Meiners. Interaction of process parameters and material properties with regard to prepreg tack in automated lay-up and draping processes. *Composites Part A: Applied Science and Manufacturing* 117 (2019): 308–316.
49. https://www.nal.res.in/en/techniques/resin-transfer-moulding-processes
50. Kong, C., H. Lee, and H. Park. Design and manufacturing of automobile hood using natural composite structure. *Composites Part B: Engineering* 91 (2016): 18–26.
51. Kong, K., B. K. Deka, S. K. Kwak, A. Oh, H. Kim, Y. B. Park, and H. W. Park. Processing and mechanical characterization of ZnO/polyester woven carbon–fiber composites with different ZnO concentrations. *Composites Part A: Applied Science and Manufacturing* 55 (2013): 152–160.
52. Nisha, M. S., P. F. Khan, and K. V. Ravali. Structural health monitoring of glass fiber reinforced polymer using nanofiber sensor. In K. Vijay Sekar, M. Gupta, A. Arockiarajan (Eds.), *Advances in Manufacturing Processes*. Springer, Singapore, 2019: pp. 245–256.
53. https://www.pinterest.com/pin/702631979334405930
54. Banerjee, A., L. Sun, S. C. Mantell, and D. Cohen. Model and experimental study of fiber motion in wet filament winding. *Composites Part A: Applied Science and Manufacturing* 29.3 (1998): 251–263.
55. Weise, B. A., K. G. Wirth, L. Völkel, M. Morgenstern, and G. Seide. Pilot-scale fabrication and analysis of graphene-nanocomposite fibers. Carbon 144 (2019): 351–361.
56. Zhao, H., X. Lan, L. Liu, Y. Liu, and J. Leng. Design and analysis of shockless smart releasing device based on shape memory polymer composites. *Composite Structures* 223 (2019): 110958.
57. Landesmann, A., C. A. Seruti, and E. D. M. Batista. Mechanical properties of glass fiber reinforced polymers members for structural applications. *Materials Research* 18.6 (2015): 1372–1383.
58. Fairuz, A. M., S. M. Sapuan, N. M. Marliana, and J. Sahari. Fabrication and effect of immersion in various solutions on mechanical properties of pultruded kenaf fiber composites: A review. In S. M. Sapuan, H. Ismail, E. S. Zainudin (Eds.), *Natural Fibre Reinforced Vinyl Ester and Vinyl Polymer Composites*. Woodhead Publishing, Duxford, 2018: pp. 109–127.

59. Bowlby, L. K., G. C. Saha, and M. T. Afzal. Flexural strength behavior in pultruded GFRP composites reinforced with high specific-surface-area biochar particles synthesized via microwave pyrolysis. *Composites Part A: Applied Science and Manufacturing* 110 (2018): 190–196.
60. Chang, B. P., W. H. Chan, M. H. Zamri, Md H. Akil, and H. G. Chuah. Investigating the effects of operational factors on wear properties of heat-treated pultruded kenaf fiber-reinforced polyester composites using Taguchi method. *Journal of Natural Fibers* 16.5 (2019): 702–717.
61. Saenz-Dominguez, I., I. Tena, A. Esnaola, M. Sarrionandia, J. Torre, and J. Aurrekoetxea. Design and characterisation of cellular composite structures for automotive crash-boxes manufactured by out of die ultraviolet cured pultrusion. *Composites Part B: Engineering* 160 (2019): 217–224.
62. Zhu, W., T. Chen, Y. Li, J. Lei, X. Chen, W. Yao, and T. Duan. High performances of artificial nacre-like graphene oxide-carrageenan bio-nanocomposite films. *Materials* 10.5 (2017): 536.
63. Vaez, M., S. Alijani, M. Omidkhah, and A. Zarringhalam Moghaddam. Synthesis, characterization and optimization of N-TiO_2/PANI nanocomposite for photodegradation of acid dye under visible light. *Polymer Composites* 39.12 (2018): 4605–4616.
64. Chen, Y. H., G. J. Zhong, Y. Wang, Z. M. Li, and L. Li. Unusual tuning of mechanical properties of isotactic polypropylene using counteraction of shear flow and β-nucleating agent on β-form nucleation. *Macromolecules* 42.12 (2009): 4343–4348.
65. Zhang, Y. F., Y. Chang, X. Li, and D. Xie. Nucleation effects of a novel nucleating agent bicyclic [2, 2, 1] heptane di-carboxylate in isotactic polypropylene. *Journal of Macromolecular Science®, Part B: Physics* 50.2 (2010): 266–274.
66. Deniz, A. E., H. A. Vural, B. Ortaç, and T. Uyar. Gold nanoparticle/polymer nanofibrous composites by laser ablation and electrospinning. *Materials Letters* 65.19–20 (2011): 2941–2943.
67. Liu, M., H. Younes, H. Hong, and G.P. Peterson. Polymer nanocomposites with improved mechanical and thermal properties by magnetically aligned carbon nanotubes. *Polymer* 166 (2019): 81–87.
68. Zhang, J., D. Liu, Q. Han, L. Jiang, H. Shao, B. Tang, W. Lei, T. Lin, and C. H. Wang. Mechanically stretchable piezoelectric polyvinylidene fluoride (PVDF)/ Boron nitride nanosheets (BNNSs) polymer nanocomposites. *Composites Part B: Engineering* 175 (2019): 107157.
69. Wang, T., Y. Li, S. Geng, C. Zhou, X. Jia, F. Yang, L. Zhang, X. Ren, and H. Yang. Preparation of flexible reduced graphene oxide/poly (vinyl alcohol) film with superior microwave absorption properties. *RSC Advances* 5.108 (2015): 88958–88964.
70. Yee, M. J., N. M. Mubarak, M. Khalid, E. C. Abdullah, and P. Jagadish. Synthesis of polyvinyl alcohol (PVA) infiltrated MWCNTs buckypaper for strain sensing application. *Scientific Reports* 8.1 (2018): 17295.
71. Wang, H., Q. Hao, X. Yang, L. Lu, and X. Wang. A nanostructured graphene/ polyaniline hybrid material for supercapacitors. *Nanoscale* 2.10 (2010): 2164–2170.
72. How, G. T. S., A. Pandikumar, H. N. Ming, and L. H. Ngee. Highly exposed {001} facets of titanium dioxide modified with reduced graphene oxide for dopamine sensing. *Scientific Reports* 4 (2014): 5044.

73. Naidek, N., K. Huang, G. Bepete, M. L. Rocco, A. Penicaud, A. J. Zarbin, and E. S. Orth. Anchoring conductive polymeric monomers on single-walled-carbon nanotubides: Towards covalently linked nanocomposites. *New Journal of Chemistry* 43 (2019): 10482.
74. Huangfu, Y., C. Liang, Y. Han, H. Qiu, P. Song, L. Wang, J. Kong, and J. Gu. Fabrication and investigation on the Fe₃O₄/thermally annealed graphene aerogel/epoxy electromagnetic interference shielding nanocomposites. *Composites Science and Technology* 169 (2019): 70–75.
75. Wang, L., H. Qiu, C. Liang, P. Song, Y. Han, Y. Han, J. Gu, J. Kong, D. Pan, and Z. Guo. Electromagnetic interference shielding MWCNT-Fe₃O₄@ Ag/epoxy nanocomposites with satisfactory thermal conductivity and high thermal stability. *Carbon* 141 (2019): 506–514.
76. Elanthamilan, E., A. Sathiyan, S. Rajkumar, E. J. Sheryl, and J. P. Merlin. Polyaniline based charcoal/Ni nanocomposite material for high performance supercapacitors. *Sustainable Energy and Fuels* 2.4 (2018): 811–819.
77. Gouthaman, A., A. Gnanaprakasam, V. M. Sivakumar, M. Thirumarimurugan, and M. A. R. Ahamed. Enhanced dye removal using polymeric nanocomposite through incorporation of Ag doped ZnO nanoparticles: Synthesis and characterization. *Journal of Hazardous Materials* 373 (2019): 493–503.
78. Pang, Y., J. Yang, T. E. Curtis, S. Luo, D. Huang, Z. Feng, J. O. Morales-Ferreiro et al. Exfoliated graphene leads to exceptional mechanical properties of polymer composite films. *ACS Nano* 13.2 (2019): 1097–1106.
79. Guo, J., J. Long, D. Ding, Q. Wang, Y. Shan, A. Umar, X. Zhang, B. L. Weeks, S. Wei, and Z. Guo. Significantly enhanced mechanical and electrical properties of epoxy nanocomposites reinforced with low loading of polyaniline nanoparticles. *RSC Advances*, 6.25 (2016): 21187–21192.
80. Liu, M., H. Younes, H. Hong, and G. P. Peterson. Polymer nanocomposites with improved mechanical and thermal properties by magnetically aligned carbon nanotubes. *Polymer* 166 (2019): 81–87.
81. Sandhya, P. K., M. S. Sreekala, M. Padmanabhan, K. Jesitha, and S. Thomas. Effect of starch reduced graphene oxide on thermal and mechanical properties of phenol formaldehyde resin nanocomposites. *Composites Part B: Engineering* 167 (2019): 83–92.
82. Alaaeddin, M. H., S. M. Sapuan, M. Y. M. Zuhri, E. S. Zainudin, and F. M. Al-Oqla. Physical and mechanical properties of polyvinylidene fluoride-short sugar palm fiber nanocomposites. *Journal of Cleaner Production* 235 (2019) 473–482.
83. Hung, P. Y., K. T. Lau, B. Fox, N. Hameed, B. Jia, and J. H. Lee. Effect of graphene oxide concentration on flexural properties of CFRP at low temperature. *Carbon* 152 (2019): 556–564.
84. Singh, K. K., S. K. Chaudhary, and R. Venugopal. Enhancement of flexural strength of glass fiber reinforced polymer laminates using multiwall carbon nanotubes. *Polymer Engineering & Science* 59.S1 (2019): E248–E261.
85. Nor, A. F. M., M. T. H. Sultan, M. Jawaid, A. M. R. Azmiand, and A. U. M. Shah. Analysing impact properties of CNT filled bamboo/glass hybrid nanocomposites through drop-weight impact testing, UWPI and compression-after-impact behaviour. *Composites Part B: Engineering* 168 (2019): 166–174.

86. Wen, X. One-pot route to graft long-chain polymer onto silica nanoparticles and its application for high-performance poly (L-lactide) nanocomposites. *RSC Advances* 9.24 (2019): 13908–13915.

87. Park, C., J. Jung, and G. J. Yun. Thermomechanical properties of mineralized nitrogen-doped carbon nanotube/polymer nanocomposites by molecular dynamics simulations. *Composites Part B: Engineering* 161 (2019): 639–650.

88. Wang, Z., Z. Jia, X. Feng, and Y. Zou. Graphene nanoplatelets/epoxy composites with excellent shear properties for construction adhesives. *Composites Part B: Engineering* 152 (2018): 311–315.

89. Liu, X., C. Li, Y. Pan, D. W. Schubert, and C. Liu. Shear-induced rheological and electrical properties of molten poly (methyl methacrylate)/carbon black nanocomposites. *Composites Part B: Engineering* 164 (2019): 37–44.

90. Bhasin, M., S. Wu, R. B. Ladani, A. J. Kinloch, C. H. Wang, and A. P. Mouritz. Increasing the fatigue resistance of epoxy nanocomposites by aligning graphene nanoplatelets. *International Journal of Fatigue* 113 (2018): 88–97.

91. Bakis, G., M. H. Kothmann, R. Zeiler, A. Brückner, M. Ziadeh, J. V. Breu, and V. Altstädt. Influence of size, aspect ratio and shear stiffness of nanoclays on the fatigue crack propagation behavior of their epoxy nanocomposites. *Polymer* 158 (2018): 372–380.

92. Bourchak, M., A. Algarni, A. Khan, and U. Khashaba. Effect of SWCNTs and graphene on the fatigue behavior of antisymmetric GFRP laminate. *Composites Science and Technology* 167 (2018): 164–173.

93. Avilés, F., A. May-Pat, M. A. López-Manchado, R. Verdejo, A. Bachmatiuk, and M. H. Rümmeli. A comparative study on the mechanical, electrical and piezoresistive properties of polymer composites using carbon nanostructures of different topology. *European Polymer Journal* 99 (2018): 394–402.

94. Bagotia, N., V. Choudhary, and D. K. Sharma. Synergistic effect of graphene/multiwalled carbon nanotube hybrid fillers on mechanical, electrical and EMI shielding properties of polycarbonate/ethylene methyl acrylate nanocomposites. *Composites Part B: Engineering* 159 (2019): 378–388.

95. Meeporn, K., and P. Thongbai. Improved dielectric properties of poly (vinylidene fluoride) polymer nanocomposites filled with Ag nanoparticles and nickelate ceramic particles. *Applied Surface Science* 481 (2018): 1160–1166.

96. Bhawal, P., S. Ganguly, T. K. Das, S. Mondal, L. Nayak, and N. C. Das. A comparative study of physico-mechanical and electrical properties of polymer-carbon nanofiber in wet and melt mixing methods. *Materials Science and Engineering: B* 245 (2018): 95–106.

97. Xiang, D., L. Wang, Y. Tang, C. Zhao, E. Harkin-Jones, and Y. Li. Effect of phase transitions on the electrical properties of polymer/carbon nanotube and polymer/graphene nanoplatelet composites with different conductive network structures. *Polymer International* 67.2 (2018): 227–235.

98. Alsharaeh, E. Polystyrene-poly (methyl methacrylate) silver nanocomposites: Significant modification of the thermal and electrical properties by microwave irradiation. *Materials* 9.6 (2016): 458.

99. Pashaei, S., S. Hosseinzadeh, and H. Hosseinzadeh. TGA investigation and morphological properties study of nanocrystalline cellulose/ag-nanoparticles nanocomposites for catalytic control of oxidative polymerization of aniline. *Polymer Composites* 40.S1 (2019): E753–E764.

100. Wang, R., C. Xie, L. Zeng, and H. Xu. Thermal decomposition behavior and kinetics of nanocomposites at low-modified ZnO content. *RSC Advances* 9.2 (2019): 790–800.
101. Terzopoulou, Z., P. A. Klonos, A. Kyritsis, A. Tziolas, A. Avgeropoulos, G. Z. Papageorgiou, and D. N. Bikiaris. Interfacial interactions, crystallization and molecular mobility in nanocomposites of Poly (lactic acid) filled with new hybrid inclusions based on graphene oxide and silica nanoparticles. *Polymer* 166 (2019): 1–12.
102. Correia, D. M., S. Ribeiro, A. da Costa, C. Ribeiro, M. Casal, S. Lanceros-Mendez, and R. Machado. Development of bio-hybrid piezoresistive nanocomposites using silk-elastin protein copolymers. *Composites Science and Technology* 172 (2019): 134–142.
103. Viskadourakis, Z., G. Perrakis, E. Symeou, J. Giapintzakis, and G. Kenanakis. Transport properties of 3D printed polymer nanocomposites for potential thermoelectric applications. *Applied Physics A* 125.3 (2019): 159.
104. Chen, J., J. Hanand, and D. Xu. Thermal and electrical properties of the epoxy nanocomposites reinforced with purified carbon nanotubes. *Materials Letters* 246 (2019): 20–23.
105. Chen, J., X. Huang, B. Sun, and P. Jiang. Highly thermally conductive yet electrically insulating polymer/boron nitride nanosheets nanocomposite films for improved thermal management capability. *ACS Nano* 13.1 (2019): 337–345.
106. Li, M., H. Zhou, Y. Zhang, Y. Liao, and H. Zhou. Effect of defects on thermal conductivity of graphene/epoxy nanocomposites. *Carbon* 130 (2018): 295–303.
107. Fang, F., S. Ran, Z. Fang, P. Song, and H. Wang. Improved flame resistance and thermo-mechanical properties of epoxy resin nanocomposites from functionalized graphene oxide via self-assembly in water. *Composites Part B: Engineering* 165 (2019): 406–416.
108. Lopez, V., A. Paton-Carrero, A. Romero, J. L. Valverde, and L. Sanchez-Silva. Improvement of the mechanical and flame-retardant properties of polyetherimide membranes modified with graphene oxide. *Polymer-Plastics Technology and Materials* 58.11 (2019): 1170–1177.
109. Lee, S., H. Min Kim, D. G. Seong, and D. Lee. Synergistic improvement of flame retardant properties of expandable graphite and multi-walled carbon nanotube reinforced intumescent polyketone nanocomposites. *Carbon* 143 (2019): 650–659.
110. Medina, L., F. Carosio, and L. A. Berglund. Recyclable nanocomposite foams of Poly (vinyl alcohol), clay and cellulose nanofibrils–Mechanical properties and flame retardancy. *Composites Science and Technology* 182 (2019): 107762.
111. Li, Y., N. Li, J. Ge, Y. Xue, W. Niu, M. Chen, Y. Du, P. X. Ma, and B. Lei. Biodegradable thermal imaging-tracked ultralong nanowire-reinforced conductive nanocomposites elastomers with intrinsical efficient antibacterial and anticancer activity for enhanced biomedical application potential. *Biomaterials* 201 (2019): 68–76.
112. Katerinopoulou, K., A. Giannakas, N. M. Barkoula, and A. Ladavos. Preparation, characterization, and biodegradability assessment of maize starch-(PVOH)/clay nanocomposite films. *Starch-Stärke* 71.1–2 (2019): 1800076.
113. Luzi, F., D. Puglia, and L. Torre. Natural fiber biodegradable composites and nanocomposites: A biomedical application. In D. Verma, E. Fortunati, S. Jain, X. Zhang (Eds.), *Biomass, Biopolymer-Based Materials, and Bioenergy*. Woodhead Publishing, Duxford, 2019: pp. 179–201.

3

Theory Behind the Improvement of Mechanical Properties through Nanofillers

3.1 Introduction

Nanotechnology and nanoscience are undoubtedly in the leading role in works related to research throughout the world. The exceptional mechanical, optical, electrical, and magnetic properties of nanoparticles are explored by investigators and/or researchers to produce a wide range of nano dimensions material to be used in a numerous range of applications. Organic nanoparticles are expected to have better potential as compared to other inorganic and metallic nanoparticles, owing to their having more flexibility and variability in tailoring and synthesis of organic compounds [1–5]. Therefore, small molecular organic nanostructured particles have the potential toward optical and electronic properties through design and optimization [6,7].

Owing to the area of the recent development in materials, inorganic nanoparticles are developing as a potential candidate due to their outstanding physical and chemical properties, which include magnetic, optical, catalytic, and electronic properties. These inorganic nanoparticles have a large surface area, high stability, variable composition, specific biological, and abundant physicochemical multifunctionality properties. In this chapter, we include an overall description of organic and inorganic nanoparticles. Organic and inorganic nanoparticles are the materials having two or more dimensions, and their range lies in between 1 and 100 nm. However, there are no clear marks that the nanoscale ranges from a biological or chemical side [8]. Usually, the nanoparticles possess exceptional physical and chemical behaviors like catalytic, magnetic, optics, electrochemical, and thermodynamic. The chemical structure, size, and shape of nanoparticles certainly have an impact on its specific properties. The nanoparticles are prepared with polymers (organic) and/or some inorganic elements (inorganic nanoparticles). Organic nanoparticles can be normally defined as solid particles comprised of organic compounds (mainly lipids or polymeric). Some common organic nanoparticles are dendrimers, liposomes, carbonaceous-based nanomaterials, and polymeric micelles. Carbon nanotubes (CNTs) come in the family of fullerenes, and their formation occurs by graphite sheets having a diameter less than 100 nm which normally are in a cylindrical shape. The shape of these CNTs may be in the form of single-layered graphite sheets or multi-layered (numerous layers of concentric graphite sheets). These CNTs possess superior strength and electrical properties and are good heat conductors. These nanotube-based nanoparticles are regularly used as biosensors as they possess a metallic or semi-conductor nature. By implication of surface functionalization, CNTs may reduce the solubility of water intake. So, these fullerene-based nanoparticles are used as drug carriers and tissue-repair scaffolds. In contrast to inorganic nanoparticles like magnetic, ceramic, metallic, quantum dots, and polystyrene nanofillers, these have a central core comprised of inorganic materials that describe their magnetic, electronic, fluorescent, and optical behaviors. Quantum dots are the class of colloidal fluorescent semi-conductor nanocrystals having a diameter about 2–10 nm. In the middle core of quantum dots there is a set of elements from groups II–VI of the periodic system (i.e., CdSe, CdTe, CdS, PbSe, ZnS, and ZnSe or III–V, i.e., GaAs, GaN, InP, and InAs), which are "overcoated" with a ZnS layer. Usually, quantum dots are photostable in nature. They depict size and compositional emission spectra and also possess a high quantum yield. They are also resistant to photo and chemical degradation. All the above salient properties make quantum dots exceptional contrast agents for imaging and labels for bioassays. Figure 3.1 reveals different types of organic and inorganic nanoparticles.

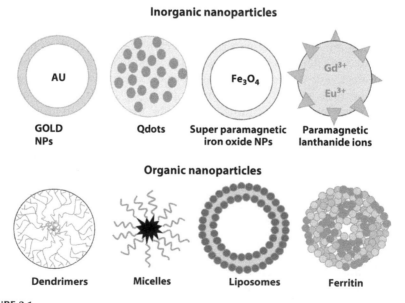

FIGURE 3.1
Different types of organic and inorganic nanoparticles. (From https://www.nanoshel.com/wp-content/uploads/2010/03/Organic-and-inorgnic-nanoparticles.jpg.)

Fiber reinforced polymeric (FRP) composites are widely used as a better unconventional advanced material in various areas such as marine, automotive, aerospace, structural applications, low temperature, and so on. The different well known advantageous properties like high specific strength and stiffness, structural integrity, performance in harsh and hostile environments can be compared to that of traditional materials [9].

But in service environments, these FRP-based composites are harshly exposed to several environmental constraints like humid atmospheres [10], low and subzero environments [11,12], high temperatures [13], thermal shock [14], thermal cycling [15], and also a combination of different types of loadings and harsh environments [16–23]. Numerous types of failures may occur in these laminated FRP composites like debonding, delamination, poor interfacial properties, out-of-plane properties, and so on. However, the interface/interphase is one of the weakest regions of FRP composites, which leads to premature failure of the composite structure in transverse loading situations. This may be attributed to transfers of the load from weak polymer matrix to strong fiber through the fiber/matrix interface/interphase. Hence, the interfacial strength is very critical and directly related to the strength and toughness of the composites [24]. The interface strength can be measured through inter-laminar shear strength and evaluated by a short beam shear test. The out-of-plane behavior of these composites may be assessed through conducting flexural tests. However, the enhancement in strength depends on the nanoparticle concentration, size, shape, and bonding between the matrix, fiber, and nanoparticles. Mostly, nanofillers are classified as organic-based [single-walled CNT (SWCNT), multiple-walled (MWCNT), graphite, and graphene], inorganic (SiO_2, SiC, etc.), metal (Fe, Al, etc.), metal oxide (TiO_2, Al_2O_3, ZnO, etc.), and others (WS_2, MoS_2 nano clay, etc.) [25]. The incorporation of nanofillers into the polymer matrix composites increases the mechanical properties, thermal stability, and lowers the permeability as compared to that of neat epoxy glass fiber reinforced polymer (GFRP) composites [26,27]. The current chapter discusses the various organic- and inorganic-based nanofillers and improvements in the mechanical properties made by incorporating the nanofillers into the polymer matrix composites. In the subsequent chapters, i.e., in Chapters 5 and 6, it is elaborately discussed about the incorporation of several nanofillers with organic and inorganic fillers.

3.2 Organic Nanofillers

In nature, the presence of organic nanofillers is abundant, and they also are used in many industrial assets. Nanofillers are comprised of atoms or molecules through self-organization and synthetic chemistry to form an extensive kind of structure such as liposomes, micelles, vesicles dendrimers, polymersomes, polymer conjugates, and polymeric and capsules nanofillers. Organic nanofillers typically include various organic or polymeric molecules that eventually lead to limitations for size control within certain ranges. Generally, these nanofillers are non-toxic and biodegradable, and some of them, such as liposomes and micelles have a hollow core as shown in Figure 3.2. Also, these nanofillers are sensitive to electromagnetic and thermal radiation, such as light and heat [28]. Furthermore, vesicles or micelles like organic nanofillers possess a dynamic character that contributes to altering the shape and size through self-assembling or fusion between the nanofillers. The abovementioned exceptional characteristics make them suitable for drug delivery systems. The drug transport capability, stability, and delivery methods of either the adsorbed drug or entrapped drug system govern their respective areas of applications and their effectiveness separated from their usual features such as the composition, size, surface morphology, etc. Organic nanofillers are extensively used in the biomedical field, like in a drug delivery system, as they are relatively well organized and also can be injected in specific parts of the body, also known as targeted drug delivery.

Dendrimers are classified as extremely branched synthetic polymers having a diameter less than 15 nm. It usually has a layered structure with a central core, different terminal groups and internal areas that define the characteristics of the dendrimer. There are multiple kinds of chemistry linked with the preparation of dendrimers, and this nature of dendrimer defines biological and soluble behavior. The properties showed by dendrimers are intrinsic in nature and normally used for tissue repair. Furthermore, dendrimers are said to be outstanding drug and image diagnosis-agent transferors by the variation of chemicals with several terminal groups.

Liposomes are generally composed of phospholipid vesicles having a diameter of about 50–100 nm with a bilayer membrane configuration the same as that of biological membranes. It may be classified according to the number of layers as multi-, oligo-, or unilamellar and also on the size. The amphiphilic behavior of liposomes helps to transfer hydrophilic drugs which are entrapped into the interior inside the membrane. The physicochemical features reveal outstanding diffusion, circulation, and penetration

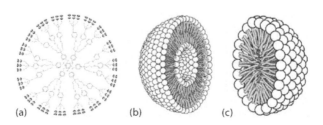

(a)   (b)   (c)

FIGURE 3.2
Organic nanofillers: (a) dendrimers, (b) liposomes and (c) micelles. (From IOP *Conf. Series: Mater. Sci. Eng.*, 263, 032019, 2017.)

properties. However, the surface of liposomes may be altered with polymers and/or ligands to enhance the drug delivery specificity [29].

3.3 Inorganic Nanofillers

Usually, inorganic nanofillers are hydrophilic, non-toxic, biocompatible, and possess good stability as compared to organic materials. These nanofillers reveal unique physical behavior when their size tends toward the nano range. Figure 3.3 shows the schematic of the surface attachment of nanofillers. The nanocrystalline quantum dots have the potential for the exceptional electronic and optical behavior that can lead to futuristic applications in the field of biomedical imaging and electro-optics. Nowadays, in several innovative and varied fields of applications, starting from magnetic to chemical behavior, the present exploration is continuously increasing toward developing the high surface to volume ratio of nanofillers as a structure of these complex nanomaterials. Basically, the structures include core/shell nanofillers and multicomponent ordered assemblies, which certainly enhance the properties. This chapter delivers the diversity of properties and structures that can be recognized in materials based on inorganic nanofillers and facts to identify the essential questions raised during controlling the properties. In addition to that, the chapter includes the different varieties, synthesis methods, and characterization procedures related to inorganic nanofillers.

3.3.1 Metal-Based Nanofillers

The nanofillers, those which are synthesized from metals to non-metric sizes either by constructive or destructive approaches, are called metal-based nanofillers. Nearly all the metals can be synthesized into their nanofillers [30].

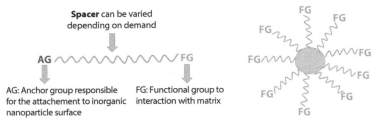

FIGURE 3.3
Schematic of surface attachment of nanofillers. (From https://www.nanoshel.com/wp-content/uploads/2010/03/Organic-and-inorgnic-nanoparticles.jpg.)

The metals usually used for nanoparticle synthesis are aluminum (Al), silver (Ag), gold (Au), cadmium (Cd), cobalt (Co), copper (Cu), iron (Fe), lead (Pb), and zinc (Zn). The different nanofillers have unique properties like sizes ranging from as low as 10 to 100 nm and surface-related phenomenon like high surface area to volume ratio, pore size, amorphous structures, surface charge and surface charge density, crystalline shapes like spherical or cylindrical, color, sensitivity, reliability, and reactivity to various environmental factors such as different temperatures (low or high), cyclic temperature variation, hygrothermal and hydrothermal, and UV irradiation.

3.3.2 Metal Oxide-Based Inorganic Nanofillers

Organic nanofillers related to metallic nanofillers have involved the material scientists can explicate to different ranges of applications. In the present exploration, these materials are being modified with different chemical functional groups that eventually enhance the chemical, optical, and mechanical properties of the material. Also, these nanofillers can be conjugated with antibodies, ligands, and drugs of interest and, hence, provide a wide range of potential applications in the field of biotechnology, magnetic separation, targeted drug delivery, drug delivery vehicles for genes, and, more importantly, diagnostic imaging. Furthermore, different imaging techniques have been established like computed tomography, magnetic resonance imaging, ultrasound, positron emission tomography, optical imaging, and surface-enhanced Raman spectroscopy as support to images for several disease states. These imaging models vary on both instrumental parts as well as in technical aspects and, more importantly, need a distinct agent with exceptional physiochemical properties. This problem brought about the development of different nanoparticulate distinct agents like magnetic nanofillers (Fe_3O_4) and silver and gold nanofillers. Furthermore, the usage of different imaging techniques with other multifunctional nanoshells and nanocages has been established [31]. From the past year's magnetic nanofillers (iron oxide), silver and gold are being used and altered to support their usage as a therapeutic and diagnostic agent [32]. Other metal oxide-based nanofillers such as Fe quickly oxidize to iron oxide (Fe_2O_3) in the presence of oxygen at room temperature that eventually raises the reactivity as compared to iron nanofillers. The synthesis of metal oxide nanofillers is mainly owed to their improved reactivity and efficiency [33]. The commonly synthesized metal oxides are aluminum oxide (Al_2O_3), titanium oxide (TiO_2), magnetite (Fe_3O_4), iron oxide (Fe_2O_3), cerium oxide (CeO_2), silicon dioxide (SiO_2), and zinc oxide (ZnO). These nanofillers have exhibited an exceptional property when compared to their respective metal equivalents.

Mahato et al. studied the flexural strength enhancement and toughening mechanisms in the case of nano-Al_2O_3-embedded glass epoxy (GE) composites. Figure 3.4 illustrates the dispersion behavior of nano-Al_2O_3 particles in the epoxy matrix of GE, 0.1 wt% and 0.5 wt% nano-Al_2O_3/GE-laminated composites. The nano-Al_2O_3 particles are mostly uniformly distributed all over the

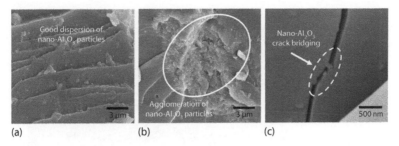

FIGURE 3.4
Dispersion of (a) 0.1 wt%, (b) 0.5 wt% nano-Al$_2$O$_3$ in GE composite, and (c) crack-bridging by nano-Al$_2$O$_3$ in 0.1 wt% nano-Al$_2$O$_3$ composite after room temperature testing. (From Mahato, K.K. et al., *Composites Part B*, 166, 688–700, 2019.)

epoxy polymer of a 0.1 wt% nano-Al$_2$O$_3$/GE composite (Figure 3.4a). In the case of 0.5 wt% nano-Al$_2$O$_3$ composites, lots of nano-Al$_2$O$_3$ particles are present as agglomerates, as illustrated in Figure 3.4b. It is known that the toughness of material could be enhanced by the mechanism known as crack-bridging [34]. The crack-bridging was observed in the fracture surface, as can be evidenced from Figure 3.4c, for the case of a 0.1 wt% nano-Al$_2$O$_3$/GE composite. Thus, it is expected that the toughness of the 0.1 wt% nano-Al$_2$O$_3$/GE composite was enhanced as compared to the other composites where the nanoparticles were present as agglomerates. Other than the 0.1 wt% nanoparticles-enhanced composite, none of the composites showed the crack-bridging phenomenon.

Nayak et al. [34] reported on the toughening mechanisms and increase in the mechanical strength of nano-TiO$_2$-filled glass fiber/epoxy composites by crack-bridging phenomenon. Figure 3.5 reveals the fracture surface of a control glass fiber and nano-composites up to 0.3 wt% TiO$_2$ in dry conditions. The field emission scanning electron microscopy (FESEM) images revealed that there is a crack-bridging by nano-TiO$_2$ particles by which the flexural and inter-laminar shear strength increases in nano-composites as compared to control glass fiber (GF) composites.

One of the potential approaches is the incorporation of various nanofillers in the case of structural FRP composites for improvement of the polymer matrix and/or interface/interphase. The addition of inorganic fillers (SiC, SiO$_2$, etc.), metal oxides (TiO$_2$, Al$_2$O$_3$, ZnO, etc.), metals (Au, Ag, Fe, Al, etc.), carbonaceous-based fillers (SWCNT, MWCNT, and graphene), and other materials (nano clay, nanosilica, MoS$_2$, WS$_2$, etc.) positively increases the mechanical properties of the polymer matrix composites [34–40].

3.3.3 Gold Nanofillers

Gold nanofillers (GNFs) are found in several sizes ranging from 2 to 100 nm; however, the particle size of about 20–50 nm exhibited the most effective cellular uptake. In terms of cell toxicity, the 40–50 nm size particles revealed

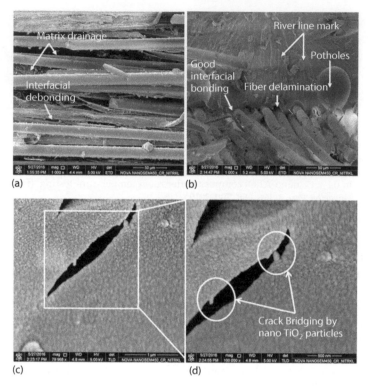

FIGURE 3.5
FESEM fracture surface features of dry samples (a) interfacial debonding and matrix drain out in 0.0 wt% TiO₂ (b) good interfacial bonding in 0.1 wt% TiO₂ (c) nano TiO₂ pull out and crack-bridging in 0.3 wt% TiO₂ (d) magnified view of nano TiO₂ pull out and crack bridging in 0.3 wt% TiO₂ as shown in (c). (From Nayak, R.K. et al., *Composites: Part A* 90, 736–74., 2016.)

better behavior. The diffusivity of 40–50 nm particles is excellent with tumors, and they easily recover it. But, on the contrary, the particles with 80–100 nm showed lesser diffusivity in tumors and resulted in accumulation of particles near the blood vessels [41,42]. The size of these different nano-fillers can be organized orderly during functionalization and the synthesis process. The size of the conjugated nanofillers is governed by the thiol/gold ratio [43]. The particle size is found to be smaller when the amount of thiol is higher [44]. The GNFs possess several advantages, such as exceptional chemical and physical properties which efficiently improve the proficiency of drugs, biocompatibility, drug loading, non-cytotoxic to the normal cells, easily reach to the targeted site with blood flow, and can be produced by several techniques [45–47]. The gold nanoshells, gold nanorods, gold nano-spheres, and gold nanocages are different kinds of GNFs [48].

The chemical, physical, and biological techniques can be established for the synthesis of GNFs. The physical procedures are primarily used to provide a

TABLE 3.1

Methods of Synthesis of GNFs

Chemical methods	Template technique
	Electrochemical technique
	PEGylation
	Turkevich technique
	Brust technique
	Perrault technique
	Martin technique
	Citrate thermal reduction technique
	Solvent free photochemical technique
	Oligonucleotide-functionalized nanofillers
	Seed-mediated technique
	Non-seed mediated technique
	Hot injection technique
	Surfactant assisted technique
	Two phase technique
	Stober process
Physical methods	Sonolysis
	γ-irradiation technique
Green methods	Green biosynthesis technique
	Sunlight irradiation technique
	New green chemistry technique

Source: Mahato, K.K. et al., *J. Polym. Environ.*, 1–27, 2017.

low yield [49]. The usage of chemical techniques in different chemical agents is to decrease metallic ions to nanofillers. In the meantime, this involves certain disadvantages, as there will be the use of toxic chemicals and evolvement of hazardous by-products [50]. In terms of medical features, in the field of nanofillers applications, the use has been increased tremendously since the biological approach for nanoparticle synthesis came into attention. These techniques are listed in (Table 3.1) [51–53].

3.3.4 Silver Nanofillers

Silver was well identified as a metal up to the recent initiation of the nanotechnology field, where it became recognized that silver may be formed and utilized at the nano level. In recent times, metallic silver has been utilized in different engineering applications, producing ultra-fine nanofillers. The nanofillers with defined particle sizes in nanometers possess typical morphologies and characteristics. These silver-based nanofillers are nowadays extensively used for cancer therapy. The AgNFs possess several advantages, such as the synthesis process can be carried out by numerous techniques and used as biosensor materials, optical properties are revealed by AgNFs, facilitate to improve wound healing; employed in the medical industry owing to their anti-bacterial, anti-viral, anti-fungal, anti-inflammatory, and osteoinductive effect [54,55].

TABLE 3.2

Methods of Synthesis of AgNFs

Chemical methods	Chemical reduction
	Micro-emulsion technology
	UV initiated photoreduction
	Photoinduced reduction
	Electrochemical method
	Electrochemical synthetic method
	Irradiation methods
	• Microwave-assisted synthesis
	• Radiolysis
	• γ-ray irradiation
	From polymers and polysaccharides Tollens' method
	Pyrolysis
Physical methods	Physical vapor condensation
	Arc-discharge method evaporation-condensation
Bio-based methods	From bacteria, fungi, yeast, algae, and plants

Source: Iravani, S. et al., *Res. Pharm. Sci.*, 9, 385–406, 2014.

3.3.4.1 Synthesis of AgNFs

The different process of synthesis of silver nanofillers can be carried out through several physical, chemical, and biological techniques. These techniques are enumerated in Table 3.2.

3.3.5 Carbon-Based

The carbon-based nanofillers are completely made up of carbon [57]. The different types of carbonaceous-based carbon nanofillers are categorized into fullerenes, graphene, CNTs, carbon nanofibers (CNFs), and carbon black, and sometimes activated carbon in nano size and are presented in Figure 3.6. According to their shapes, carbon nanofillers are characterized within three groups; they are: (a) spherical fillers or zero-dimensional; (b) cylindrical particles or one-dimensional; and (c) two-dimensional particles. The illustrations for each of these three different groups are, respectively: (a) nano-diamond; (b) CNT and CNF; and (c) graphene nanoplatelet. The previous investigations have shown that the addition of carbon-based NPs such as CNTs, CNFs, and graphene nanoplatelets in various polymers enhanced numerous properties (mechanical, thermal, and electrical) as compared to the unmodified polymers [58–61].

3.3.6 Fullerenes

Fullerenes constitute carbon molecules of spherical shape and are composed of carbon atoms bonded together by SP^2 hybridization. Near about 28–1500 carbon atoms are arranged in order to form spherical structures

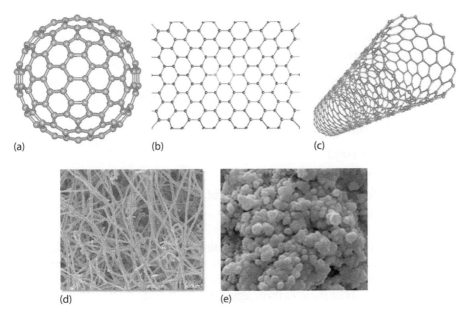

FIGURE 3.6
Carbonaceous-based nanofillers: (a) fullerenes, (b) graphene, (c) carbon nanotubes, (d) carbon nanofibers, and (e) carbon black.

with diameters nearly 8.2 nm for single-layered and 4 to 36 nm for multi-layered fullerenes. It is a pure carbon molecule made with a minimum of 60 carbon atoms. The shape of a fullerene is quite similar to that of a geodesic dome or a soccer ball. It is occasionally mentioned that the buck ball is named after the discoverer of the geodesic dome Buckminster Fuller, and the fullerene is more formally named after him. Fullerenes are seen as favorable constituents of future micro-electromechanical systems and in the world of nanotechnology. Fullerenes comprise 12 pentagonal and 20 hexagonal rings as the foundation of the icosahedral symmetry closed cage structure. It is sp^2 hybridized, and each carbon atom is bonded to three others. The electronic bonding and geodesic factors in the structure provide the stability of the molecule. In theory, an infinite number of fullerenes can exist, their structure based on pentagonal and hexagonal rings [62]. Fullerenes have numerous advantages like high tensile strength, high ductility, high electrical conductivity, relative chemical inactivity, and high resistance to heat [63].

Physical properties of C60 (fullerene) [64]

- Density: 1.65 g cm^{-3}
- Standard heat of formation: 9.08 kcal mol^{-1}
- Index of refraction: 2.2 (600 nm)

- Boiling point: sublimes at 800 K
- Resistivity: 1014 Ohm m^{-1}
- Vapor density: N/A
- Crystal form: hexagonal cubic
- Vapor pressure: 5×10^{-6} torr at room temperature

Organoleptic properties

- Color: black solid
- Odor: odorless
- Buckyball soot: very finely divided black powder
- Fullerite: brown/black powder

Generally, in some common solvents, fullerenes are starts to solubles such as benzene, toluene, or chloroform. When fullerene powder shakes up with toluene and filters the mixture, a red solution is obtained. As the solvent evaporates, crystals of pure carbon appear.

Production/synthesis of fullerenes [65]

- Laser vaporization of carbon in an inert atmosphere
- Resistive heating of graphite
- Inductive heating of graphite or another source (acetylene, etc.)
- Pyrolysis of hydrocarbon (naphthalene)
- Total synthesis of fullerene

3.3.7 Graphene

Graphene is one of the allotropes of carbon. The structure of graphene is a hexagonal network of a honeycomb lattice made up of carbon atoms in a two-dimensional planar surface. Usually, the graphene is found in the form of graphene sheets and the thickness of the sheets is around 1 nm. Graphene is a two-dimensional, highly conjugated and single-atom layered carbon nanomaterial, whose exploration history can be traced back to 1859. The word graphene was created as a combination of graphite and the suffix-ene by Hanns-Peter Boehm, who pioneered single-layer carbon foils in 1962 [66,67]. The exclusive geometry plane and structure of monolayer graphene provide its excellent properties, including high Young's modulus (~1100 GPa), excellent electrical (~106 S/cm) and thermal conductivity (~5000 W/mK), high fracture strength (~125 GPa), fast mobility of charge carriers (~200,000 cm^2/Vs), and a large specific surface area (theoretically calculated value, 2630 m^2/g) [68,69]. Graphene is one of the evolving carbon-based materials having excellent electrical, mechanical thermal conductivity, high specific surface area, and is corrosion-resistant. It also finds applications in the fields of electronics, energy storage and conversion, biotechnology, and especially improvement in composite fiber materials. Due to the abovementioned properties, the usage of graphene as nanofillers is on high demand by researchers. When fibrous polymeric matrix composites are modified with graphene, fibers show a number of improved or newer properties such as anti-bacteria, hydrophobicity adsorption performance, and conductivity, which are advantageous for a wide range of applications. There are several ways to prepare graphene, but the most common way is the reduction of graphene oxide, exfoliation of graphite, and chemical vapor deposition. However, one main difficulty that needs to be corrected for graphene before its applications is the lower dispersibility in common organic and inorganic solvents. The good dispersion of graphene in some solvents is an essential change to the formation of homogeneous nano-composites. Therefore, the change of graphene to modify its solubility is serious for wide-ranging commercial applications. Graphene can typically be altered with covalent and non-covalent methods. The non-covalent procedures comprise p–p stacking interactions, hydrogen bonding, electrostatic interaction, van der Waals force, and coordination bonds. This alteration process can extremely preserve graphene's natural structure, however, the interactions between functionalities and the graphene surface are relatively weak; therefore, it is not appropriate for a few applications where strong interactions are present. Graphene is used to strengthen the properties of fibrous-reinforced polymer composite materials that have been examined by several material scientists and researchers [39,70,71].

3.3.8 Carbon Nanotubes

CNTs are composed of graphene nanofoil having a honeycomb lattice of carbon atoms that is twisted into hollow cylinder shapes to form nanotubes of such diameter for a single-layer as low as 0.7 nm and for a multi-layered CNT about 100 nm, with the length changing from a few micrometers to several millimeters. The ends of the CNT can either be closed or hollow by a half fullerene molecule. CNTs are nanostructures resulting from rolled graphene sheets, having excellent mechanical, chemical, and physical behavior. Since the discovery of CNTs by using high-resolution transmission electron microscopy, researchers have been motivated for extreme experimental- and theoretical-based investigations on carbon nanotubes for use in various modes of applications. CNTs are allotropes of carbon with a nanostructure that can have a length-to-diameter ratio greater than 1,000,000 [72,73]. Since the innovation of CNTs, they have been widely used as nanofillers for an extensive range of applications including structural, chemical, electrochemical, electrical, electronics, and biological. The ultra-strong (~500 GPa) and modulus (~1 TPa) CNTs, with a very low density (~1.5 g/cc), enormous high specific surface area (~10–20 m^2/g), and electrical conductivity (~106 S/m), are one of the prime choices for reinforcing soft polymeric materials [74]. Certainly, a CNT is one of the most favorable nanofillers for the proper improvement of the mechanical, thermal, and electrical behavior of polymer matrix composites [75–77]. The incorporation of 1 wt.% of MWCNT into a carbon fiber/epoxy composite has revealed the substantial improvement in its interlaminar shear strength and tensile behavior with a reduction in dispersity [78]. At lower and cryogenic temperatures, the strength of an epoxy has been stated to be considerably improved due to the incorporation of CNTs [79]. As temperature-influenced damages further may damage the properties of the composite material, its structural reliability must be certified at the service temperature [80]. The existence of MWCNT in the glass fiber/epoxy composite considerably blocked its water absorption tendency at a lower aging temperature (25°C) [81]. However, MWCNT reinforcement in a GE composite adversely affected its high-temperature water resistance due to the generation of unfavorable thermal and hygroscopic stresses at the MWCNT/polymer interfaces [81].

CNTs can be conjugated with several biological molecules with proteins, drugs, and nucleic acid to give bio-functionalities [82,83]. In addition to that, the aromatic ring existing on the CNT surface permits effective loading of aromatic molecules, such as chemotherapeutic drugs via stacking [84]. Mostly, CNTs exist in two forms: (a) SWCNT and (b) MWCNT structures. Due to the outstanding properties, such as high-aspect ratio, ultra-lightweight, high specific strength and stiffness, high thermal conductivity, and remarkable electronic properties ranging from metallic to semi-conducting materials, CNTs find usage in several applications [37,85]. In a biologically related field, SWCNTs show a photoluminescence behavior that could be expertly useful in diagnostics, while MWCNTs show a wider surface that allows a more efficient internal encapsulation and external functionalization with active molecules. They have been both used for varied roles with biosensors, field-effect transistors, and scanning probe elements [86].

Figure 3.7 reveals the significance of various factors like CNT shape, size, dispersion, interaction, alignment, and so on which shows the main role in the fabrication of CNT-reinforced polymeric composites. By using numerous methods, results can be enhanced. With a combination of the excellent properties of the composite synthesis route and the filler, these two deliver an exceptional foundation of composite materials for several engineering applications. These composites cover a wide range of applications as conductive glue, gas storage devices, energy storage devices, sensors, lightweight aircraft applications, defense, and so on [87–89]. As a CNT possesses small dimensions and exhibit strong van der Waals forces, they have the affinity to stick together [90]. As a result, there is inefficient transfer of stress from the matrix to the fillers. To avoid these problems, functionalization methods are used, i.e., covalent and non-covalent functionalization. The addition of CNTs into the polymer matrix results in substantial changes in the mechanical,

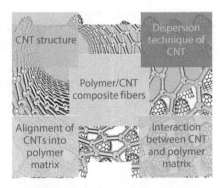

FIGURE 3.7
Major factors affecting the development of CNT/polymer nano-composites during processing. (From Mittal, G. et al., *J. Ind. Eng. Chem.*, 21, 11–25, 2015.)

electrical, and thermal properties of the polymer composite [38,91,92]. Furthermore, the chapter discusses in short detail about the importance and use of CNTs synthesis.

3.3.8.1 Single-Walled Carbon Nanotubes

SWCNTs are comprised of a single cylindrical carbon layer with a diameter in the range of 0.4–2 nm, subject to the synthesis temperature. It was seen and reported that the higher the temperature, the larger is the diameter of CNTs [93]. The structure of SWCNTs may be a helical, chiral, armed chair, and zigzag arrangements [94]. Furthermore, to the strength enhancement, CNT nanofillers have been proved to be advantageous for the enhancement in the toughness of polymeric-based materials. The easiest technique to see the enhancement in the toughness of polymer phase by CNT nanofiller in the presence of a substantial amount of CNTs ahead of the generation of crack tip, which delays the propagation of crack. However, this mechanism is the same for all the nanofiller-embedded polymeric composites. Furthermore, some more methods of toughening mechanisms are present, which are exceptional for CNTs due to their dimensions and geometry.

Figure 3.8 shows the SEM images of fractured surfaces neat epoxy and CNT-enhanced epoxy composites. The presence of river line markings in Figure 3.8a shows the brittle nature of the epoxy polymer. However, in Figure 3.8b, the image shows the rather rough ductile failure of the polymeric composite [95]. The other method of toughness in CNT is the crack bridging as observed by Qian et al. [91] by in situ deformation of an MWCNT/polystyrene nano-composite film in a TEM. The CNTs are aligned perpendicular to the crack direction where they are obstructing the crack propagation by crack bridging phenomenon.

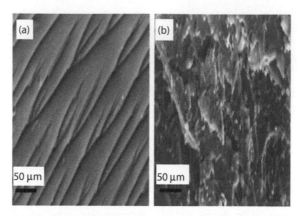

FIGURE 3.8
SEM micrographs of quasi-statically fractured surfaces: (a) epoxy, (b) epoxy–CNT. (From Jajam, K.C. et al., *Compos. Part Appl. Sci. Manuf.*, 59, 57–69, 2014.)

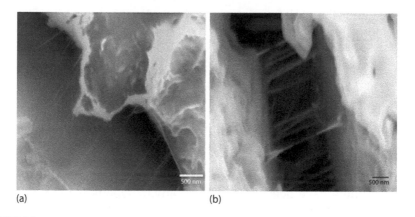

(a)   (b)

FIGURE 3.9
SEM-micrographs of DWCNT($-NH_2$)/epoxy composites. A crack, induced by etching, is bridged by (a) amino-functionalized and (b) non-functionalized double-wall carbon nanotubes. (From Gojny, F.H. et al., *Compos. Sci. Technol.*, 65, 2300–2313, 2005.)

A similar kind of CNT polymer crack-bridging was observed by Gojny et al. [96]. The high strain to failure of CNTs [97,98] relative to the epoxy permits an enormous deformation of the nanotubes, which is several times that of the epoxy matrix. This also allows a crack-bridging of functionalized double walled carbon nanotubes (DWCNTs) (Figure 3.9a), as well as non-functionalized DWCNTs in the epoxy (3.9(b)), as observed by SEM.

Rathore et al. [37] also reported good dispersion, agglomeration, and crack-bridging of CNT nanofillers in the epoxy polymer. Figure 3.10 shows the good dispersion state of MWCNTs in the polymer of 0.1% and 0.5% MWCNT/epoxy glass fiber-laminated composites. The MWCNTs are generally isolated from each other and homogeneously distributed throughout the matrix of 0.1% MWCNT/epoxy glass fiber composite (Figure 3.10a), whereas

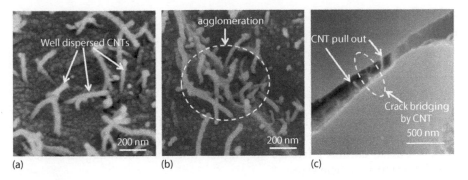

FIGURE 3.10
Dispersion of (a) 0.1%, (b) 0.5% MWCNT in GE composite, and (c) CNT pullout and crack-bridging by CNT in 0.1% MWCNT–GE composite after room temperature testing. (From Rathore, D.K. et al., *Compos. Part Appl. Sci. Manuf.*, 84, 364–376, 2016.)

in 0.5% MWCNT/epoxy glass fiber composite, local bunches of MWCNTs are found which form agglomerates as shown in Figure 3.10b. The toughness mechanism can be seen through the increment in glass fiber composite strength due to the incorporation of 0.1% MWCNT, this may be attributed to the nanotube pullout and crack-bridging by nanotubes as can be seen from Figure 3.10c.

3.3.8.2 Multiple-Walled Carbon Nanotubes

The structural arrangement of MWCNTs is of numerous coaxial cylinders, each made of a single graphene sheet surrounding a hollow cone. The outer diameter of MWCNTs lies between 2 and 100 nm, while the inner diameter is ranging between 1 and 3 nm, and their overall length varies from one to several micrometers. MWCNTs nanoparticle structures can be divided into two types based on their arrangements of graphite layers: (a) parchment-like structure, which comprises of a graphene sheet rolled up around it and (b) the Russian doll model, where a layer of the graphene sheet is arranged within a concentric structure [99].

3.3.8.3 Methods of CNTs Synthesis [100–102]

- Plasma-based synthesis method or arc discharge evaporation method
- Laser ablation method
- Thermal synthesis process
- Gas-phase methods
- Chemical vapor deposition
- Plasma-enhanced chemical vapor deposition

3.3.9 Carbon Nanofiber

For the synthesis of CNF, the same graphene nanofoil is used as that of CNT, but twisted into a cup or cone shape in place of regular cylindrical tubes.

3.3.10 Carbon Black

Carbon black is an amorphous material made up of carbon, normally spherical in shape and has diameters ranging from 20 to 70 nm. The bond energies between these carbon particles are higher, and they form agglomerates of around 500 nm particle size.

3.3.11 Quantum Dots

When a solid reveals a discrete distinction of optical and electronic properties having variation in particle size less than 100 nm, it is called a nanostructure. It is divided as: (a) one-dimensional, e.g., quantum wires, (b) two-dimensional, e.g., thin films or quantum wells, and (c) zero-dimensional or dots [103]. A crystal having a nanometer dimension is often called a quantum dot QD. Normally, QD sizes lie in the range of 2–20 nm [104]. However, for better properties, diameter of QDs should be less than 10 nm [105,106]. The type of materials used during the synthesis of QDs contributes to its dimensions [107].

Quantum dots provide a controlled size, and due to this, it is possible to have control over the conductive behavior of the materials. Normally, QDs are important in the field of optical applications owing to their high extinction coefficient. It is always beneficial to tune the size of QDs for several applications. Usually, the smaller dots allow having the advantage of better quantum effects. Whereas, the larger QDs provide greater spectrum shift, as compared to smaller dots exhibiting less prominent quantum effects. QDs are zero-dimensional in nature, possessing a sharper density of states than higher-dimensional structures. This results in excellent optical and transport behavior [108]. The different properties of quantum dots that make it useful in many applications are highly resistant to photobleaching and extremely high brightness when excited.

3.3.12 Silica Nanofillers

The usage of silica nanofillers in nano-composites has been an active area of exploration in research by many investigators in order to have some improved properties in mechanical, thermal, as well as fracture toughness [109–115]. This results in the improvement of yield strength and elastic modulus of the composite [114,116,117]. Silica NPs are also used in biomedical field-related applications. It is here divided as mesoporous or non-porous (solid) NPs, and both of which showed amorphous silica structures. Mesoporous silica NPs are categorized by the mesopores having a 2–50 nm pore size and were extensively used for the delivery of active payloads based on chemical or physical adsorption [118,119]. Furthermore, silica-based NPs are used for enhanced drug delivery systems. Multifunctional silica NPs have the target of nanolevel nano-composites with better chemical, magnetic, and optical behavior in a combined single nanostructure [120,121]. Numerous techniques are available for the synthesis of silica NPs, such as the sol-gel method, reverse micro-emulsion, Stober method, spray drying, laser ablation, template method, hydrolysis, thermal vaporization, thermal annealing, heating degradation, polycondensation, soft templating and hard templating, polymer beads templating, vesicle templating, micelle templating, and micro-emulsion templating [122,123].

Nanosilica is being considered as a promising potential material and is presently found in several nano-composite applications, industrial constructions, and used as adhesives materials. In several cases, the addition of nanosilica has strengthened, stiffened, and toughened the polymer matrix system. Investigations have shown that the use of nanosilica as fillers causes a further substantial enhancement in the stiffness, better than microfillers [124,125].

References

1. Link, S., M. B. Mohamed, and M. A. El-Sayed. Simulation of the optical absorption spectra of gold nanorods as a function of their aspect ratio and the effect of the medium dielectric constant. *The Journal of Physical Chemistry B* 103 (1999): 3073–3077. doi:10.1021/jp990183f.
2. Peng, X., L. Manna, W. Yang, J. Wickham, E. Scher, and A. Kadavanich, and A. P. Alivisatos. Shape control of CdSe nanocrystals. *Nature* 404 (2000): 59–61. doi:10.1038/35003535.
3. Kasai, H., S. Okada, and H. Nakanishi. Polydiacetylene microcrystals and their third-order optical nonlinearity. In F. Kajzar, M.V. Agranovich (Eds.), *Multiphoton Light Driven Multielectron Process Organ New Phenom Materials Applied*, Springer, the Netherlands, 2000: pp. 345–356.
4. Jana, N. R., L. Gearheart, and C. J. Murphy. Wet chemical synthesis of silver nanorods and nanowires of controllable aspect ratio. *Chemical Communications* (2001): 617–618. doi:10.1039/B100521I.
5. Murphy, C. J., and N. R. Jana. Controlling the aspect ratio of inorganic nanorods and nanowires. *Advanced Materials* 14 (2002): 80–82. doi:10.1002/1521-4095(20020104)14:1< 80::AID-ADMA80>3.0.CO;2-#.
6. Fu, H., D. Xiao, J. Yao, and G. Yang. Nanofibers of 1,3-diphenyl-2-pyrazoline induced by cetyltrimethylammonium bromide micelles. *Angewandte Chemie International Edition in English* 42 (2003): 2883–2886. doi:10.1002/anie.200350961.
7. Nguyen, T.-Q., R. Martel, P. Avouris, M. L. Bushey, L. Brus, and C. Nuckolls. Molecular interactions in one-dimensional organic nanostructures. *Journal of the American Chemical Society* 126 (2004): 5234–5242. doi:10.1021/ja031600b.

8. Boverhof, D. R., C. M. Bramante, J. H. Butala, S. F. Clancy, M. Lafranconi, J. West, and S. C. Gordon. Comparative assessment of nanomaterial definitions and safety evaluation considerations. *Regulatory Toxicology and Pharmacology* 73 (2015): 137–150. doi:10.1016/j.yrtph.2015.06.001.
9. Mangalgiri, P. D. Composite materials for aerospace applications. *Bulletin of Materials Science* 22 (1999): 657–664. doi:10.1007/BF02749982.
10. Ray, B. C. Temperature effect during humid ageing on interfaces of glass and carbon fibers reinforced epoxy composites. *Journal of Colloid and Interface Science* 298 (2006): 111–117. doi:10.1016/j.jcis.2005.12.023.
11. Shukla, M. J., D. S. Kumar, D. K. Rathore, R. K. Prusty, and B. C. Ray. An assessment of flexural performance of liquid nitrogen conditioned glass/epoxy composites with multiwalled carbon nanotube. *Journal of Composite Materials* 50 (2016): 3077–3088. doi:10.1177/0021998315615648.
12. Ray, B. C. Effects of thermal and cryogenic conditionings on mechanical behavior of thermally shocked glass fiber-epoxy composites. *Journal of Reinforced Plastics and Composites* 24 (2005): 713–717. doi:10.1177/0731684405046081.
13. Cao, S., Z. Wu, and X. Wang. Tensile properties of CFRP and hybrid FRP composites at elevated temperatures. *Journal of Composite Materials* 43 (2009): 315–330. doi:10.1177/0021998308099224.
14. Ray, B. C. Thermal shock on interfacial adhesion of thermally conditioned glass fiber/epoxy composites. *Materials Letters* 58 (2004): 2175–2177. doi:10.1016/j.matlet.2004.01.035.
15. Ghasemi, A. R., and M. Moradi. Effect of thermal cycling and open-hole size on mechanical properties of polymer matrix composites. *Polymer Testing* 59 (2017): 20–28. doi:10.1016/j.polymertesting.2017.01.013.
16. Mahato, K. K., A. O. Fulmali, R. Kattaguri, K. Dutta, R. K. Prusty, and B. C. Ray. Effect of severely thermal shocked MWCNT enhanced glass fiber reinforced polymer composite: An emphasis on tensile and thermal responses. *IOP Conference Series: Materials Science and Engineering* 338 (2018): 012057. doi:10.1088/1757-899X/338/1/012057.
17. Mahato, K. K., K. Dutta, and B. C. Ray. Static and dynamic behavior of fibrous polymeric composite materials at different environmental conditions. *Journal of Polymers and the Environment* (2017): 1–27. doi:10.1007/s10924-017-1001-x.
18. Li, G., J. Wu, and W. Ge. Effect of loading rate and chemical corrosion on the mechanical properties of large diameter glass/basalt-glass FRP bars. *Thermal and Cryogenic Conditionings on Mechanical* 93 (2015): 1059–1066. doi:10.1016/j.conbuildmat.2015.05.044.
19. Mahato, K. K., D. K. Rathore, K. Dutta, and B. C. Ray. Effect of loading rates of severely thermal-shocked glass fiber/epoxy composites. *Composites Communications* 3 (2017): 7–10. doi:10.1016/j.coco.2016.11.001.
20. Ray, B. Ch. Loading rate effects on mechanical properties of polymer composites at ultralow temperatures. *Journal of Applied Polymer Science* 100 (2006): 2289–2292. doi:10.1002/app.22853.
21. Sethi, S., D. K. Rathore, and B. C. Ray. Effects of temperature and loading speed on interface-dominated strength in fibre/polymer composites: An evaluation for in-situ environment. *Materials & Design* 65 (2015): 617–626. doi:10.1016/j.matdes.2014.09.053.

22. Koller, R., S. Chang, and Y. Xi. Fiber-reinforced polymer bars under freeze-thaw cycles and different loading rates. *Journal of Composite Materials* 41 (2007): 5–25. doi:10.1177/0021998306063154.
23. Prusty, R. K., D. K. Rathore, B. P. Singh, S. C. Mohanty, K. K. Mahato, and B. C. Ray. Experimental optimization of flexural behaviour through inter-ply fibre hybridization in FRP composite. *Construction and Building Materials* 118 (2016): 327–336. doi:10.1016/j.conbuildmat.2016.05.054.
24. Dirand, X., B. Hilaire, J. P. Soulier, and M. Nardin. Interfacial shear strength in glass-fiber/vinylester-resin composites. *Composites Science and Technology* 56 (1996): 533–539. doi:10.1016/0266-3538(96)00040-1.
25. Naffakh, M., A. M. Díez-Pascual, C. Marco, G. J. Ellis, and M. A. Gómez-Fatou. Opportunities and challenges in the use of inorganic fullerene-like nanoparticles to produce advanced polymer nanocomposites. *Progress in Polymer Science* 38 (2013): 1163–1231. doi:10.1016/j.progpolymsci.2013.04.001.
26. Anjana, R., and K. E. George. Reinforcing effect of nano kaolin clay on PP/HDPE blends. *Int. J. Eng. Res. Ind. Appl.* 2 (2012): 868–872.
27. Garc, M., W. E. van Zyl, and B. Boukamp. Polypropylene/Sio$_2$ Nanocomposites with Improved Mechanical Properties, *Rev. Adv. Mater. Sci.* 6 (2004): 169–217.
28. Sen, P. Application of nanoparticles in waste water treatment, (n.d.). https://www.academia.edu/13968188/Application_of_Nanoparticles_in_Waste_Water_Treatment (Accessed July 8, 2019).
29. Anu Mary Ealia, S., and M. P. Saravanakumar. A review on the classification, characterisation, synthesis of nanoparticles and their application. *IOP Conference Series: Materials Science and Engineering* 263 (2017): 032019. doi:10.1088/1757-899X/263/3/032019.
30. Salavati-Niasari, M., F. Davar, and N. Mir. Synthesis and characterization of metallic copper nanoparticles via thermal decomposition. *Polyhedron* 27 (2008): 3514–3518. doi:10.1016/j.poly.2008.08.020.
31. Mody, V. V., R. Siwale, A. Singh, and H. R. Mody. Introduction to metallic nanoparticles. *Journal of Pharmacy and Bioallied Sciences* 2 (2010): 282. doi:10.4103/0975-7406.72127.
32. Moghimi, S. M., A. C. Hunter, and J. C. Murray. Nanomedicine: Current status and future prospects. *The FASEB Journal: Official Publication of the Federation of American Societies for Experimental Biology* 19 (2005): 311–330. doi:10.1096/fj.04-2747rev.
33. Tai, C. Y., C.-T. Tai, M.-H. Chang, and H.-S. Liu. Synthesis of magnesium hydroxide and oxide nanoparticles using a spinning disk reactor. *Industrial & Engineering Chemistry Research* 46 (2007): 5536–5541. doi:10.1021/ie060869b.
34. Nayak, K. K. Mahato, and B. C. Ray. Water absorption behavior, mechanical and thermal properties of nano TiO$_2$ enhanced glass fiber reinforced polymer composites. *Composites Part A: Applied Science and Manufacturing* 90 (2016): 736–747. doi:10.1016/j.compositesa.2016.09.003.
35. Rodriguez, A. J., M. E. Guzman, C.-S. Lim, and B. Minaie. Mechanical properties of carbon nanofiber/fiber-reinforced hierarchical polymer composites manufactured with multiscale-reinforcement fabrics. *Carbon* 49 (2011): 937–948. doi:10.1016/j.carbon.2010.10.057.

36. Nayak, R. K., K. K. Mahato, B. C. Routara, and B. C. Ray. Evaluation of mechanical properties of Al_2O_3 and TiO_2 nano filled enhanced glass fiber reinforced polymer composites. *Journal of Applied Polymer Science* 133 (2016). doi:10.1002/app.44274.
37. Rathore, D. K., R. K. Prusty, D. S. Kumar, and B. C. Ray. Mechanical performance of CNT-filled glass fiber/epoxy composite in in-situ elevated temperature environments emphasizing the role of CNT content. *Composites Part A: Applied Science and Manufacturing* 84 (2016): 364–376. doi:10.1016/j.compositesa.2016.02.020.
38. Allaoui, A., S. Bai, H. M. Cheng, and J. B. Bai. Mechanical and electrical properties of a MWNT/epoxy composite. *Composites Science and Technology* 62 (2002): 1993–1998. doi:10.1016/S0266-3538(02)00129-X.
39. Ji, X., Y. Xu, W. Zhang, L. Cui, and J. Liu. Review of functionalization, structure and properties of graphene/polymer composite fibers. *Composites Part A: Applied Science and Manufacturing* 87 (2016): 29–45. doi:10.1016/j.compositesa.2016.04.011.
40. Zhou, Y., F. Pervin, L. Lewis, and S. Jeelani. Fabrication and characterization of carbon/epoxy composites mixed with multi-walled carbon nanotubes. *Materials Science and Engineering: A* 475 (2008): 157–165. doi:10.1016/j.msea.2007.04.043.
41. El-Sayed, I. H., X. Huang, and M. A. El-Sayed. Selective laser photo-thermal therapy of epithelial carcinoma using anti-EGFR antibody conjugated gold nanoparticles. *Cancer Letter* 239 (2006): 129–135. doi:10.1016/j.canlet.2005.07.035.
42. Jadzinsky, P. D., G. Calero, C. J. Ackerson, D. A. Bushnell, and R. D. Kornberg. Structure of a thiol monolayer-protected gold nanoparticle at 1.1 A resolution. *Science* 318 (2007): 430–433. doi:10.1126/science.1148624.
43. Bhattacharya, S., and A. Srivastava. Synthesis of gold nanoparticles stabilised by metal-chelator and the controlled formation of close-packed aggregates by them. *Journal of Chemical Sciences* 115 (2003): 613–619. doi:10.1007/BF02708252.
44. Li, L., M. Fan, R. C. Brown, J. (Hans) V. Leeuwen, J. Wang, W. Wang, Y. Song, and P. Zhang. Synthesis, properties, and environmental applications of nanoscale iron-based materials: A review. *Critical Reviews in Environmental Science and Technology* 36 (2006): 405–431. doi:10.1080/10643380600620387.
45. Chithrani, D. B., S. Jelveh, F. Jalali, M. van Prooijen, C. Allen, R. G. Bristow, R. P. Hill, and D. A. Jaffray. Gold nanoparticles as radiation sensitizers in cancer therapy. *Journal of Radiation Research* 173 (2010): 719–728. doi:10.1667/RR1984.1.
46. Lan, M.-Y., Y.-B. Hsu, C.-H. Hsu, C.-Y. Ho, J.-C. Lin, and S.-W. Lee. Induction of apoptosis by high-dose gold nanoparticles in nasopharyngeal carcinoma cells. *Auris Nasus Larynx* 40 (2013): 563–568. doi:10.1016/j.anl.2013.04.011.
47. Khan, A. K., R. Rashid, G. Murtaza, and A. Zahra. Gold nanoparticles: Synthesis and applications in drug delivery. *Tropical Journal of Pharmaceutical Research* 13 (2014): 1169–1177. doi:10.4314/tjpr.v13i7.23.
48. Han, G., C. T. Martin, and V. M. Rotello. Stability of gold nanoparticle-bound DNA toward biological, physical, and chemical agents. *Chemical Biology & Drug Design* 67 (2006): 78–82. doi:10.1111/j.1747-0285.2005.00324.x.
49. Malik, M. A., P. O'Brien, and N. Revaprasadu. A simple route to the synthesis of core/shell nanoparticles of chalcogenides. *Chemistry of Materials* 14 (2002): 2004–2010. doi:10.1021/cm011154w.
50. Sriram, M. I., S. B. M. Kanth, K. Kalishwaralal, and S. Gurunathan. Antitumor activity of silver nanoparticles in Dalton's lymphoma ascites tumor model. *International Journal of Nanomedicine* 5 (2010): 753–762. doi:10.2147/IJN.S11727.

51. Ankamwar, B. Biosynthesis of gold nanoparticles (Green-gold) using leaf extract of terminalia catappa. *Journal of Chemical* (2010). doi:10.1155/2010/745120.
52. Heidari, Z., R. Sariri, and M. Salouti. Gold nanorods-bombesin conjugate as a potential targeted imaging agent for detection of breast cancer. *Journal of Photochemistry and Photobiology B* 130 (2014): 40–46. doi:10.1016/j.jphotobiol.2013.10.019.
53. Madhusudhan, A., G. B. Reddy, M. Venkatesham, G. Veerabhadram, D. A. Kumar, S. Natarajan, M.-Y. Yang, A. Hu, and S. S. Singh. Efficient pH dependent drug delivery to target cancer cells by gold nanoparticles capped with carboxymethyl chitosan. *International Journal of Molecular Sciences* 15 (2014): 8216–8234. doi:10.3390/ijms15058216.
54. Pulit, J., M. Banach, R. Szczygłowska, and M. Bryk. Nanosilver against fungi silver nanoparticles as an effective biocidal factor. *Acta Biochimica Polonica* 60 (2013): 795–798.
55. Qu, D., W. Sun, Y. Chen, J. Zhou, and C. Liu. Synthesis and in vitro antineoplastic evaluation of silver nanoparticles mediated by Agrimoniae herba extract. *International Journal of Nanomedicine* 9 (2014): 1871–1882. doi:10.2147/IJN.S58732.
56. Iravani, S., H. Korbekandi, and S. V. Mirmohammadi, and B. Zolfaghari. Synthesis of silver nanoparticles: Chemical, physical and biological methods. *Research in Pharmaceutical Sciences* 9 (2014): 385–406.
57. Bhaviripudi, S., E. Mile, S. A. Steiner, A. T. Zare, M. S. Dresselhaus, A. M. Belcher, and J. Kong. CVD synthesis of single-walled carbon nanotubes from gold nanoparticle catalysts. *Journal of the American Chemical Society* 129 (2007): 1516–1517. doi:10.1021/ja0673332.
58. Alishahi, E., S. Shadlou, S. Doagou-R, and M. R. Ayatollahi. Effects of carbon nanoreinforcements of different shapes on the mechanical properties of epoxy-based nanocomposites. *Macromolecular Materials and Engineering* 298 (2013): 670–678. doi:10.1002/mame.201200123.
59. Ayatollahi, M. R., S. Shadlou, M. M. Shokrieh, and M. Chitsazzadeh. Effect of multi-walled carbon nanotube aspect ratio on mechanical and electrical properties of epoxy-based nanocomposites. *Polymer Testing* 30 (2011): 548–556. doi:10.1016/j.polymertesting.2011.04.008.
60. Choi, Y. K., K. I. Sugimoto, S. M. Song, and M. Endo. Mechanical and thermal properties of vapor-grown carbon nanofiber and polycarbonate composite sheets. *Materials Letters* 59 (2005): 3514–3520. doi:10.1016/j.matlet.2005.05.082.
61. Gao, J., W. Li, H. Shi, M. Hu, and R. K. Y. Li. Preparation, morphology, and mechanical properties of carbon nanotube anchored polymer nanofiber composite. *Composites Science and Technology* 92 (2014): 95–102. doi:10.1016/j.compscitech.2013.12.008.
62. Hummelen, J. C., B. W. Knight, F. Lepeq, F. Wudl, J. Yao, and C. L. Wilkins. Preparation and characterization of fulleroid and methanofullerene derivatives. *The Journal of Organic Chemistry* 60 (1995): 532–538.
63. Bakry, R., R. M. Vallant, M. Najam-ul-Haq, M. Rainer, Z. Szabo, C. W. Huck, and G. K. Bonn. Medicinal applications of fullerenes. *International Journal of Nanomedicine* 2 (2007): 639–649.
64. Morgan, G. J. Historical review: Viruses, crystals and geodesic domes. *Trends in Biochemical Sciences* 28 (2003): 86–90.
65. Shanbogh, P. P., and N. G. Sundaram. Fullerenes revisited. *Resonance* 20 (2015): 123–135. doi:10.1007/s12045-015-0160-0.

66. Van Noorden, R. Production: Beyond sticky tape. *Nature* 483 (2012): S32–S33. doi:10.1038/483S32a.
67. Boehm, H.-P. Graphene—How a laboratory curiosity suddenly became extremely interesting. *Angewandte Chemie International Edition* 49 (2010): 9332–9335. doi:10.1002/anie.201004096.
68. Pan, Y., N. G. Sahoo, and L. Li. The application of graphene oxide in drug delivery. *Expert Opinion on Drug Delivery* 9 (2012): 1365–1376. doi:10.1517/17425247.2012.729575.
69. Service, R. F. Materials science. Carbon sheets an atom thick give rise to graphene dreams. *Science* 324 (2009): 875–877. doi:10.1126/science.324_875.
70. Ren, G., Z. Zhang, X. Zhu, B. Ge, F. Guo, X. Men, and W. Liu. Influence of functional graphene as filler on the tribological behaviors of Nomex fabric/phenolic composite. *Composites Part A: Applied Science and Manufacturing* 49 (2013): 157–164. doi:10.1016/j.compositesa.2013.03.001.
71. Xu, Z., and C. Gao. In situ polymerization approach to graphene-reinforced nylon-6 composites. *Macromolecules* 43 (2010): 6716–6723. doi:10.1021/ma1009337.
72. Wang, N., K. K. Fung, W. Lu, and S. Yang. Structural characterization of carbon nanotubes and nanoparticles by high-resolution electron microscopy. *Chemical Physics Letters* 229 (1994): 587–592. doi:10.1016/0009-2614(94)01114-1.
73. Zhang, B., Q. Chen, H. Tang, Q. Xie, M. Ma, L. Tan, Y. Zhang, and S. Yao. Characterization of and biomolecule immobilization on the biocompatible multi-walled carbon nanotubes generated by functionalization with polyamidoamine dendrimers. *Colloids Surface B Biointerfaces* 80 (2010): 18–25. doi:10.1016/j.colsurfb.2010.05.023.
74. Sahoo, N. G., S. Rana, J. W. Cho, L. Li, and S. H. Chan. Polymer nanocomposites based on functionalized carbon nanotubes. *Progress in Polymer Science* 35 (2010): 837–867. doi:10.1016/j.progpolymsci.2010.03.002.
75. Lau, A. K.-T., and D. Hui. The revolutionary creation of new advanced materials—Carbon nanotube composites. *Composites Part B: Engineering* 33 (2002): 263–277. doi:10.1016/S1359-8368(02)00012-4.
76. Gojny, F. H., M. H. G. Wichmann, B. Fiedler, W. Bauhofer, and K. Schulte. Influence of nano-modification on the mechanical and electrical properties of conventional fibre-reinforced composites. *Composites Part A: Applied Science and Manufacturing* 36 (2005): 1525–1535. doi:10.1016/j.compositesa.2005.02.007.
77. Li, Q., M. Zaiser, J. R. Blackford, C. Jeffree, Y. He, and V. Koutsos. Mechanical properties and microstructure of single-wall carbon nanotube/elastomeric epoxy composites with block copolymers. *Materials Letters* 125 (2014): 116–119. doi:10.1016/j.matlet.2014.03.096.
78. Zhang, J., S. Ju, D. Jiang, and H.-X. Peng. Reducing dispersity of mechanical properties of carbon fiber/epoxy composites by introducing multi-walled carbon nanotubes. *Composites Part B: Engineering* 54 (2013): 371–376. doi:10.1016/j.compositesb.2013.05.046.
79. Chen, Z.-K., J.-P. Yang, Q.-Q. Ni, S.-Y. Fu, and Y.-G. Huang. Reinforcement of epoxy resins with multi-walled carbon nanotubes for enhancing cryogenic mechanical properties. *Polymer* 50 (2009): 4753–4759. doi:10.1016/j.polymer.2009.08.001.

80. Sethi, S., and B. C. Ray. Environmental effects on fibre reinforced polymeric composites: Evolving reasons and remarks on interfacial strength and stability. *Advances in Colloid and Interface Science* 217 (2015): 43–67. doi:10.1016/j.cis.2014.12.005.

81. Prusty, R. K., D. K. Rathore, and B. C. Ray. Water-induced degradations in MWCNT embedded glass fiber/epoxy composites: An emphasis on aging temperature. *Journal of Applied Polymer Science* 135 (2018): 45987. doi:10.1002/app.45987.

82. McDevitt, M. R., D. Chattopadhyay, B. J. Kappel, J. S. Jaggi, S. R. Schiffman, C. Antczak, J. T. Njardarson, R. Brentjens, and D. A. Scheinberg. Tumor targeting with antibody-functionalized, radiolabeled carbon nanotubes. *The Journal of Nuclear Medicine Official Publication Society Nuclear Medicine* 48 (2007): 1180–1189. doi:10.2967/jnumed.106.039131.

83. Liu, Z., S. M. Tabakman, Z. Chen, and H. Dai. Preparation of carbon nanotube bioconjugates for biomedical applications. *Nature Protocols* 4 (2009): 1372–1382. doi:10.1038/nprot.2009.146.

84. Liu, Z., X. Sun, N. Nakayama-Ratchford, and H. Dai. Supramolecular chemistry on water-soluble carbon nanotubes for drug loading and delivery. *ACS Nano* 1 (2007): 50–56. doi:10.1021/nn700040t.

85. Jin, H., D. A. Heller, and M. S. Strano. Single-particle tracking of endocytosis and exocytosis of single-walled carbon nanotubes in NIH-3T3 cells. *Nano Letter* 8 (2008): 1577–1585. doi:10.1021/nl072969s.

86. Feazell, R. P., N. Nakayama-Ratchford, H. Dai, and S. J. Lippard. Soluble single-walled carbon nanotubes as longboat delivery systems for platinum (IV) anticancer drug design. *Journal of the American Chemical Society* 129 (2007): 8438–8439. doi:10.1021/ja073231f.

87. Kurahatti, R. V., A. O. Surendranathan, S. A. Kori, N. Singh, A. V. R. Kumar, and S. Srivastava. Defence applications of polymer nanocomposites. *Defence Science Journal* 60 (2010): 551–563. doi:10.14429/dsj.60.578.

88. M. F. L. De Volder, S. H. Tawfick, R. H. Baughman, and A. J. Hart. Carbon nanotubes: Present and future commercial applications. *Science* 339 (2013): 535–539. doi:10.1126/science.1222453.

89. Ma, Y., W. Cheung, D. Wei, A. Bogozi, P. L. Chiu, L. Wang, F. Pontoriero, R. Mendelsohn, and H. He. Improved conductivity of carbon nanotube networks by in situ polymerization of a thin skin of conducting polymer. *ACS Nano* 2 (2008): 1197–1204. doi:10.1021/nn800201n.

90. Mittal, G., V. Dhand, K. Y. Rhee, S.-J. Park, and W. R. Lee. A review on carbon nanotubes and graphene as fillers in reinforced polymer nanocomposites. *Journal of Industrial and Engineering Chemistry* 21 (2015): 11–25. doi:10.1016/j.jiec.2014.03.022.

91. Qian, D., E. C. Dickey, R. Andrews, and T. Rantell. Load transfer and deformation mechanisms in carbon nanotube-polystyrene composites. *Applied Physics Letters* 76 (2000): 2868–2870. doi:10.1063/1.126500.

92. Biercuk, M. J., M. C. Llaguno, M. Radosavljevic, J. K. Hyun, A. T. Johnson, and J. E. Fischer. Carbon nanotube composites for thermal management. *Applied Physics Letters* 80 (2002): 2767–2769. doi:10.1063/1.1469696.

93. Klumpp, C., K. Kostarelos, M. Prato, and A. Bianco. Functionalized carbon nanotubes as emerging nanovectors for the delivery of therapeutics. *Biochimica et Biophysica Acta—Biomembranes* 1758 (2006): 404–412. doi:10.1016/j.bbamem.2005.10.008.

94. Danailov, D., P. Keblinski, S. Nayak, and P. M. Ajayan. Bending properties of carbon nanotubes encapsulating solid nanowires. *Journal of Nanoscience and Nanotechnology* 2 (2002): 503–507.

95. Jajam, K. C., M. M. Rahman, M. V. Hosur, and H. V. Tippur. Fracture behavior of epoxy nanocomposites modified with polyol diluent and amino-functionalized multi-walled carbon nanotubes: A loading rate study. *Composites Part A: Applied Science and Manufacturing* 59 (2014): 57–69. doi:10.1016/j.compositesa.2013.12.014.

96. Gojny, F. H., M. H. G. Wichmann, B. Fiedler, and K. Schulte. Influence of different carbon nanotubes on the mechanical properties of epoxy matrix composites—A comparative study. *Composites Science and Technology* 65 (2005): 2300–2313. doi:10.1016/j.compscitech.2005.04.021.

97. Yu, M.-F., O. Lourie, M. J. Dyer, K. Moloni, T. F. Kelly, and R. S. Ruoff. Strength and breaking mechanism of multiwalled carbon nanotubes under tensile load. *Science* 287 (2000): 637–640. doi:10.1126/science.287.5453.637.

98. Yu, M. F., Files, B. S., Arepalli, S., and Ruoff, R. S. Tensile loading of ropes of single wall carbon nanotubes and their mechanical properties. *Physical Review Letters* 84 (2000): 5552–5555. doi:10.1103/PhysRevLett.84.5552.

99. Zhang, S., K. Yang, and Z. Liu. Carbon nanotubes for in vivo cancer nanotechnology. *Science China Chemistry* 53 (2010): 2217–2225. doi:10.1007/s11426-010-4115-8.

100. Prasek, J., J. Drbohlavova, J. Chomoucka, J. Hubalek, O. Jasek, V. Adam, and R. Kizek. Methods for carbon nanotubes synthesis—review. *Journal of Materials Chemistry* 21 (2011): 15872–15884. doi:10.1039/C1JM12254A.

101. Varshney, K. Carbon nanotubes: A review on synthesis, properties and applications, *International Journal of Engineering Research and General Science*, 2, (2014).

102. Shin, U. S., I.-K. Yoon, G.-S. Lee, W.-C. Jang, J. C. Knowles, and H.-W. Kim. Carbon nanotubes in nanocomposites and hybrids with hydroxyapatite for bone replacements. *Journal of Tissue Engineering* 2011 (2011). doi:10.4061/2011/674287.

103. Chang, Y.-P., F. Pinaud, J. Antelman, and S. Weiss. Tracking bio-molecules in live cells using quantum dots. *Journal of Biophotonics* 1 (2008): 287–298. doi:10.1002/jbio.200810029.

104. Klusoň, P., M. Drobek, H. Bártková, and I. Budil. Welcome in the Nanoworld. *Chemické Listy* 101 (2007). http://www.chemicke-listy.cz/ojs3/index.php/chemicke-listy/article/view/1796 (Accessed July 9, 2019).

105. Král, V., J. Šotola, P. Neuwirth, Z. Kejík, K. Záruba, and P. Martásek. Nanomedicine—current status and perspectives: A big potential or just a catchword? *Chemické Listy* 100 (2006). http://www.chemicke-listy.cz/ojs3/index.php/chemicke-listy/article/view/1957 (Accessed July 9, 2019).

106. Ferancová, A., and J. Labuda. DNA biosensors based on nanostructured materials. (2008). doi:10.1002/9783527621507.ch11.

107. Fujioka, K., M. Hiruoka, K. Sato, N. Manabe, R. Miyasaka, S. Hanada, A. Hoshino, R. D. et al. Luminescent passive-oxidized silicon quantum dots as biological staining labels and their cytotoxicity effects at high concentration. *Nanotechnology* 19 (2008): 415102. doi:10.1088/0957-4484/19/41/415102.

108. Maiti, A., and S. Bhattacharyya. Review: Quantum dots and application in medical science, *International Journal of Chemistry and Chemical Engineering* 3 (2013): 37–42
109. Brunner, A. J., A. Necola, M. Rees, Ph. Gasser, X. Kornmann, R. Thomann, and M. Barbezat. The influence of silicate-based nano-filler on the fracture toughness of epoxy resin. *Engineering Fracture Mechanics* 73 (2006): 2336–2345. doi:10.1016/j.engfracmech.2006.05.004.
110. Jordan, J., K. I. Jacob, R. Tannenbaum, M. A. Sharaf, and I. Jasiuk. Experimental trends in polymer nanocomposites—A review. *Materials Science and Engineering: A* 393 (2005): 1–11. doi:10.1016/j.msea.2004.09.044.
111. Han, J. T., and K. Cho. Layered silicate-induced enhancement of fracture toughness of epoxy molding compounds over a wide temperature range. *Macromolecular Materials and Engineering* 290 (2005): 1184–1191. doi:10.1002/mame.200500195.
112. Rosso, P., L. Ye, K. Friedrich, and S. Sprenger. A toughened epoxy resin by silica nanoparticle reinforcement. *Journal of Applied Polymer Science* 100 (2006): 1849–1855. doi:10.1002/app.22805.
113. Zhang, H., Z. Zhang, K. Friedrich, and C. Eger. Property improvements of in situ epoxy nanocomposites with reduced interparticle distance at high nanosilica content. *Acta Material* 54 (2006): 1833–1842. doi:10.1016/j.actamat.2005.12.009.
114. Ragosta, G., M. Abbate, P. Musto, G. Scarinzi, and L. Mascia. Epoxy-silica particulate nanocomposites: Chemical interactions, reinforcement and fracture toughness. *Polymer* 46 (2005): 10506–10516. doi:10.1016/j.polymer.2005.08.028.
115. Ou, C.-F., and M.-C. Shiu. Epoxy composites reinforced by different size silica nanoparticles. *Journal of Applied Polymer Science* 115 (2010): 2648–2653. doi:10.1002/app.29809.
116. Deng, S., L. Ye, and K. Friedrich. Fracture behaviours of epoxy nanocomposites with nano-silica at low and elevated temperatures. *Journal of Materials Science* 42 (2007): 2766–2774. doi:10.1007/s10853-006-1420-x.
117. Battistella, M., M. Cascione, B. Fiedler, M. H. G. Wichmann, M. Quaresimin, and K. Schulte. Fracture behaviour of fumed silica/epoxy nanocomposites. *Composites Part A: Applied Science and Manufacturing* 39 (2008): 1851–1858. doi:10.1016/j.compositesa.2008.09.010.
118. Vallet-Regí, M., F. Balas, and D. Arcos. Mesoporous materials for drug delivery. *Angewandte Chemie International Edition* 46 (2007): 7548–7558. doi:10.1002/anie.200604488.
119. Slowing, I. I., J. L. Vivero-Escoto, C.-W. Wu, and V. S.-Y. Lin. Mesoporous silica nanoparticles as controlled release drug delivery and gene transfection carriers. *Advanced Drug Delivery Reviews* 60 (2008): 1278–1288. doi:10.1016/j.addr.2008.03.012.
120. Barbé, C., J. Bartlett, L. Kong, K. Finnie, H. Q. Lin, M. Larkin, S. Calleja, A. Bush, and G. Calleja. Silica particles: A novel drug-delivery system. *Advanced Materials* 16 (2004): 1959–1966. doi:10.1002/adma.200400771.
121. Wu, J., Z. Ye, G. Wang, and J. Yuan. Multifunctional nanoparticles possessing magnetic, long-lived fluorescence and bio-affinity properties for time-resolved fluorescence cell imaging. *Talanta* 72 (2007): 1693–1697. doi:10.1016/j.talanta.2007.03.018.

122. Yang, P., S. Gai, and J. Lin. Functionalized mesoporous silica materials for controlled drug delivery. *Chemical Society Reviews* 41 (2012): 3679–3698. doi:10.1039/c2cs15308d.
123. Tang, L., and J. Cheng. Nonporous silica nanoparticles for nanomedicine application. *Nano Today* 8 (2013): 290–312. doi:10.1016/j.nantod.2013.04.007.
124. Liang, Y. L., and R. A. Pearson. Toughening mechanisms in epoxy–silica nanocomposites (ESNs). *Polymer* 50 (2009): 4895–4905. doi:10.1016/j.polymer.2009.08.014.
125. Ruiz-Pérez, L., G. J. Royston, J. P. A. Fairclough, and A. J. Ryan. Toughening by nanostructure. *Polymer* 49 (2008): 4475–4488. doi:10.1016/j.polymer.2008.07.048.

4

Hydrothermal Behavior of Polymer Nanocomposites

4.1 Diffusion Theory of Moisture Absorption

4.1.1 Fick's Model

In 1855, a German physiologist Adolph Fick developed a mathematical model, which was later named Fick's law of diffusion. Fick's first law of diffusion states that for an isotropic medium, the rate of diffusion through any cross-section is directly proportional to the concentration gradient normal to it and quantitatively represented as:

$$F = -D\left(\frac{\partial C}{\partial x}\right) \tag{4.1}$$

where F is the mass flux and D is the diffusivity. Fick's second law is also known as the law of diffusion. The law of diffusion depends on Fick's first law.

$$\frac{\partial C}{\partial t} = F\left(-\frac{\partial}{\partial x}\right) \tag{4.2}$$

$$\frac{\partial C}{\partial t} = \left(-\frac{\partial}{\partial x}\right) \times \left(-D\left(\frac{\partial C}{\partial x}\right)\right) \tag{4.3}$$

$$\frac{\partial C}{\partial t} = \frac{\partial(D\partial C)}{\partial x(\partial x)}. \tag{4.4}$$

Moisture concentration is independent of diffusivity so:

$$\frac{\partial C}{\partial t} = D\frac{\partial^2 C}{\partial x^2}. \tag{4.5}$$

The water absorption by the fiber-reinforced polymer composite, Fick's equation further simplified. The water absorbed by the composite (M) at the time (t) is determined by gravimetric analysis is given by:

$$M = \frac{\text{Weight of material with moisture content} - \text{Weight of dry material}}{\text{Weight of dry material}}.$$

$$(4.6)$$

If diffusion is carried out at a constant temperature and relative humidity, the water content in a thin flat plate kind of sample is given by:

$$M = G(M_\infty - M_i) + M_i. \qquad (4.7)$$

This expression can be used during absorption and desorption. Where M is the moisture content in the material, M_∞ is the water content at saturation point, M_i is the moisture content at the initial condition, and G is the time-dependent parameter:

$$G = 1 - \frac{8}{\pi^2} \sum_{j=1}^{\infty} \frac{\exp\left[-(2j+1)^2 \pi^2 \left(\frac{D_z t}{s^2}\right)\right]}{(2j+1)^2}. \qquad (4.8)$$

Figure 4.1 shows the different methods of moisture diffusion, where s depends on the thickness of plate (h) and D_z is the diffusivity in the direction perpendicular to the plane of the plate. If the diffusion occurs from both parallel sides, then $s = h$, and if one side is restricted by the mean of some insulation or water coating, then $s = 2h$.

Figure 4.2 shows exact and approximate solution of Equation 4.8 [1].

$$G = 1 - \exp\left[-7.3\left(\frac{D_z t}{s^2}\right)^{0.75}\right]. \qquad (4.9)$$

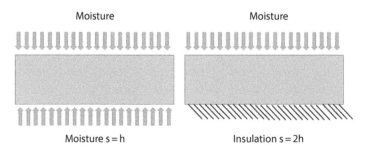

Moisture   Moisture

Moisture s = h   Insulation s = 2h

FIGURE 4.1
A schematic representation of the different process.

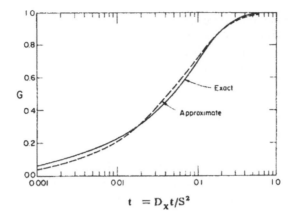

FIGURE 4.2
A comparison between exact and approximate solution. (From Shen, C.H. and Springer, G.S., *J. Comp. Mater.*, 10, 2–20, 1976.)

For the unidirectional fiber-reinforced polymer composite material, with the orientation of fibers at angles α, β, and γ with respect to the x-, y-, and z-directions, respectively, the diffusion coefficient can be determined by:

$$D_z = D_{11} \cos^2 \gamma + D_{22} \sin^2 \gamma. \tag{4.10}$$

D_{11} represents the diffusivity in the longitudinal direction, and D_{22} represents the diffusivity in the transverse direction of the fiber axis. However, it has been common practice to determine the diffusion coefficient by plotting the parameter M versus $\sqrt{t}$, as shown in Figure 4.3.

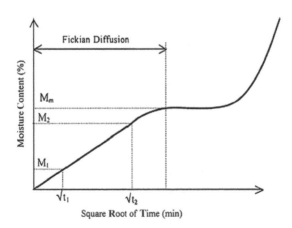

FIGURE 4.3
Water absorption behavior of the fiber-reinforced polymer composite. (From Nayak, R.K. et al., *Compos. Part A App. Sci. Manuf.*, 90, 736–747, 2016.)

In the beginning, the water uptake process is linear between M and $\sqrt{t}$, as the diffusion coefficient is independent of the water content. The slope of the linear portion of the curve carries the information regarding the diffusivity, as per the expression given below:

$$\text{slope} = \frac{M_2 - M_1}{\sqrt{t_2} - \sqrt{t_1}} = \frac{4M_\infty}{h\sqrt{\pi}}\sqrt{D_{eff}}. \tag{4.11}$$

In the case when the water is not absorbed from the edge of the sample or for an infinitely large flat sample, the diffusivity (D_z) becomes equal to the D_{eff}. However, considering the diffusion from edges, the diffusivity can be determined by taking the sample dimensions, which are length (L), width (w), and thickness.

$$D_z = D_{eff}\left(1 + \frac{h}{L} + \frac{h}{w}\right)^{-2}. \tag{4.12}$$

The moisture absorbed by the fiber-reinforced polymer composite follows Fick's law of diffusion. From various research work, it is observed that composites absorbed the water molecules particularly at the initial linear absorption region of Fick's model. Most of the researchers observed that the moisture absorbed by the fiber-reinforced polymer composites follows Fick's law of diffusion. It is also observed that some of the fiber reinforced polymer (FRP) composite materials do not follow Fick's law of moisture absorption.

4.1.2 Langmuirian Model

The Langmuirian diffusion model is a time-varying diffusion coefficient model, and it is also known as the dual sorption model or the two-phase diffusion model. The Langmuirian model assumed that the absorbed water molecules are divided into two parts. The first one is the preferential absorption of the diffusivity water molecules into the free phase of the material with diffusivity D_γ. Subsequently, the water molecules are constrained to the polymeric chains of the resin with a probability γ and the probability of becoming unbound is β. From the above assumption, Carter and Kibler derived a non-Fickian model for water adsorption. The approximation model is expressed below.

$$M = M_\infty \left[\frac{\beta}{\beta = \gamma} \exp(-\gamma t)\left(1 - \frac{8}{\pi^2}\sum_{i=1}^{\infty(odd)} \frac{\exp(-ki^2t)}{i^2}\right)\right.$$

$$\left. + \frac{\beta}{\beta + \gamma}\left(\exp(-\beta t) - \exp(-\gamma t) + (1 - \exp(-\beta t))\right)\right]; \; 2\gamma, 2\beta << k \text{ and } k = \frac{\pi^2 D_\gamma}{h^2}. \tag{4.13}$$

For a short time duration, the absorption equation can be modified as:

$$M \approx M_\infty \frac{4}{\pi^{3/2}}\left(\frac{\beta}{\beta+\gamma}\right)\sqrt{kt}; \quad 2\gamma, 2\beta << k; t \le 0.7k. \tag{4.14}$$

For a long time duration the equation can be written as:

$$M \approx M_\infty\left(1-\frac{\gamma}{\beta+\gamma}\exp(-\beta t)\right); \quad t \ge \frac{1}{k}. \tag{4.15}$$

Many researchers have adopted this Langmuirian diffusion model for the absorption of water molecules by fiber-reinforced polymer composite materials. Figure 4.4 shows the modified Langmuirian model, which is known as the Carter-Kibler model and more precisely fits with a glass/epoxy composite than Fick's model [3].

4.1.3 Hindered Model

Let us assume that the diffusion occurred in one dimension (1D), at any point z and time t, the Langmuirian diffusion follows the differential equation mentioned below.

$$D\frac{\partial^2 n}{\partial z^2} = \frac{\partial n}{\partial t} + \frac{\partial N}{\partial t} \tag{4.16}$$

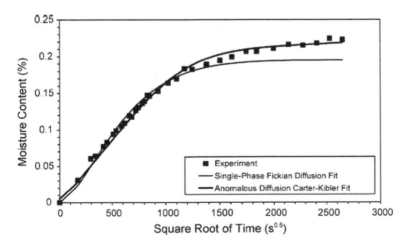

FIGURE 4.4
Moisture absorption by glass/epoxy by Fick's and Carter-Kibler. (From Kumosa, et al., *Compos. Part A: App. Sci. Manuf.*, 35, 1049–1063, 2004.)

$$\frac{\partial N}{\partial t} = \gamma n - \beta N. \tag{4.17}$$

The n represents the number of free molecules per unit volume, and N represents the number of bounded molecules per unit volume. The process of absorption continues until the number of bounded molecules, which are becoming mobile, becomes equal to the corresponding number of mobile molecules, which are becoming bound per unit time, thus the equilibrium is established when the following condition is met.

$$\gamma n = \beta N. \tag{4.18}$$

The water molecules are absorbed by the face exposed to the humid environment, which is initial at a dry condition as is given by.

$$M = M_\infty \left[1 - \frac{8}{\pi^2} \sum_{i-1}^{\infty(odd)} \left(\frac{r_i^+ \exp\left(-r_i^- t\right) - r_i^- \exp\left(-r_i^+ t\right)}{i^2 \left(r_i^+ - r_i^-\right)} \right) \right]$$

$$+ \frac{8}{\pi^2} \left(\frac{k\beta}{\beta + \gamma} \right) \sum_{i-1}^{\infty(odd)} \left(\frac{r_i^+ \exp\left(-r_i^- t\right) - \exp\left(-r_i^+ t\right)}{\left(r_i^+ - r_i^-\right)} \right) \tag{4.19}$$

$$r_i^\pm = \frac{1}{2} \left[\left(ki^2 + \beta + \gamma\right) \pm \sqrt{\left(ki^2 + \beta + \gamma\right)^2 - 4k\beta i^2} \right]. \tag{4.20}$$

This one-dimensional hindered diffusion model can be extended into 3D by making some assumptions. The diffusion can take place from all directions simultaneously, due to the diffusing molecules each taking the other into account during the interactions.

$$D_x^* \frac{\partial^2 n^*}{\partial \left(x^*\right)^2} + D_y^* \frac{\partial^2 n^*}{\partial \left(y^*\right)^2} + D_z^* \frac{\partial^2 n^*}{\partial \left(z^*\right)^2} = \mu \frac{\partial n^*}{\partial t^*} + \left(1 - \mu\right)\left(n^* - N^*\right), \tag{4.21}$$

where

$$n^* = \frac{n}{n_\infty}; N^* = \frac{N}{N_\infty}; t^* = \beta t$$

$$x^* = \frac{x}{l}; y^* = \frac{y}{w}; z^* = \frac{z}{h}$$

$$D_x^* = \frac{D_x}{l^2 \left(\beta + \gamma\right)}; D_y^* = \frac{D_y}{w^2 \left(\beta + \gamma\right)}; D_z^* = \frac{D_z}{h^2 \left(\beta + \gamma\right)}; \mu = \frac{\beta}{\beta + \gamma}.$$

μ is a hindrance coefficient which is dimensionally less quantity, and takes care of the temporal and spatial water concentration. The value of μ is depending on the rate of change in concentration of free molecules, its values vary between 0 and 1. μ also governs the relationship between n^* and N^*, at equilibrium, both of these parameters approach unity. It is also indicative of the dependence of β and γ with the diffusion process. In the case of hindrance, the coefficient becomes 1, the 3D hindrance diffusion model expressed in Equation 4.22 becomes the same as the 3D Fickian diffusion model as mentioned below:

$$D_x^* \frac{\partial^2 n^*}{\partial \left(x^* \right)^2} + D_y^* \frac{\partial^2 n^*}{\partial \left(y^* \right)^2} + D_z^* \frac{\partial^2 n^*}{\partial \left(z^* \right)^2} = \mu \frac{\partial n^*}{\partial t^*}. \tag{4.22}$$

Determining the amount of moisture in equilibrium with a set of discrete parameters in a polymer combination is of practical importance. Conventionally, this is confirmed by the practical saturation value. However, defining saturation is crucial. The American Society for Testing and Materials (ASTM) recommends that the D5229 be saturated, with three consecutive gravimetric analyses showing no more than 0.01%. In the case of polymeric composites, another important term known as "pseudoquilibrium" refers to the rapid change in moisture absorption rate, as shown in Figure 4.5.

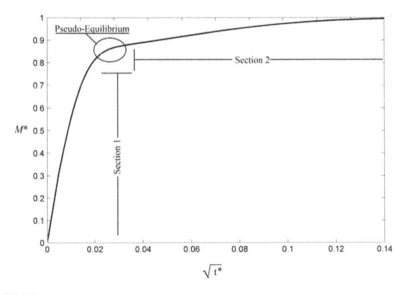

FIGURE 4.5
Schematic representation of both sections of hindered diffusion. (From Grace, L.R. and Altan, M.C., *Compos. Part A App. Sci. Manuf.*, 43, 1187–1196, 2012.)

Both Fick's and non-Fick's absorption models create linear relationships between M and $\sqrt{t}$ during the early stages of dispersion, which is before the pseudoquiliberium takes place.

However, the difference between Fick's and non-Fick's models is apparent after this episode of false balance. For the Fick's model, the moisture content does not increase during the later stages of dispersal, indicating that the equilibrium is equal to the false equilibrium. In the case of hindered diffusion model (HDM), the moisture content continues to increase after the false equilibrium, but to a lesser extent. Thus, accurately predicting equilibrium materials in practice is a difficult and time-consuming process.

Tang et al. [4] reported the dynamic behavior of carbon fiber reinforced polymer (CFRP) samples over distilled water. They compared the experimental data with the Fick's scattering model. An ideal fit can be seen at the beginning of water absorption. At the same time, Fick's model avoids experimental data exposure. Grace et al. [3] provided the same experimental data with HDM, and it is interesting that a longer propagation period such as Figure 4.6. can be better fitted.

But the shorter experimental time frame limits the accurate and precise prediction of the equilibrium moisture content, that is M_∞. The lowest value of M (1.85%) corresponds to a 4.5-year saturation period, while the highest value (3.87%) represents a 16-year saturation period. This implies that short-term gravity water absorption analysis is insufficient to predict the exact amount of water in equilibrium. In later stages of HDM, slower growth rates take longer for better accuracy.

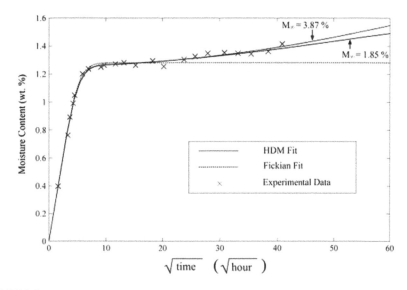

FIGURE 4.6
Application of a hindered model. (From Grace, L.R. and Altan, M.C., *Compos. Part A App. Sci. Manuf.*, 43, 1187–1196, 2012.)

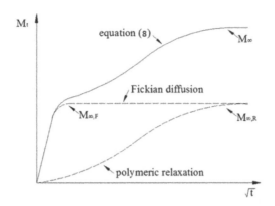

FIGURE 4.7
Theoretical absorption curve due to Fickian diffusion and polymeric relaxation. (From Jiang, X. et al., *Compos. Part B-Eng.*, 45, 407–416, 2013.)

4.1.4 Dual-Stage Model

The two-phase diffusion model is based on the assumption that both Fickian and non-Fickian diffusion occur simultaneously during moisture dissipation in polymer composites. As shown in Figure 4.7, the different stages of distribution are characterized by the formation of a dominant spreading mechanism.

In the initial phase of diffusion, the Fickian model prevails. As the non-Fickian model prevails over time, the relaxation ratio increases, and the rate of diffusion slows down. If the moisture content of Fickian and relaxation behavior is represented by M_F and M_R respectively, pure water content (M) can be considered as the dominant position of both behaviors.

$$M = M_F + M_R \qquad (4.23)$$

During this period, when the relaxation process is complete, saturation will be involved. The Fickian contribution has been rewritten from Equation 4.9, and the contribution of the relaxation process can be distinguished from the following approach:

$$M_R = M_{\infty,R}\left[1-\exp(-kt)\right], \qquad (4.24)$$

where $M_{\infty,R}$ and k represent the saturation moisture content for the relaxation mechanism and relaxation rate constant, respectively. Thus, Equation 4.23 becomes:

$$M = M_{\infty,F}\left[1-\exp\left\{-7.3\left(\frac{Dt}{h^2}\right)^{0.75}\right\}\right] + M_{\infty,R}\left[1-\exp(-kt)\right]. \qquad (4.25)$$

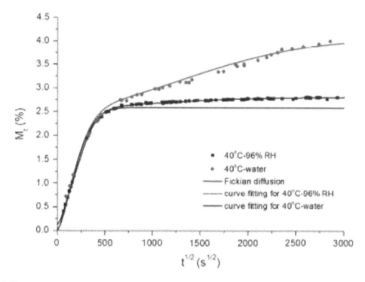

FIGURE 4.8
A comparison between Fick's model and dual stage model. (From Jiang, X. et al., *Compos. Part B-Eng.*, 45, 407–416, 2013.)

Several articles have shown that polymer composites follow dual-phase diffusion behavior. Jiang et al. [5] reported that for experimental polyurethane adhesions, the experimental kinetic data are better fitted, as shown in Figure 4.8. The Fickian's adaptation at two different relative humidity (RH) levels, with no significant difference at the same temperature, indicates that a Fickian model is less sensitive to a RH difference. As can be seen from Figure 4.8, there is a clear deviation in the kinetic curves due to the different RH levels after the linear state. For water conditioning, moisture is much higher due to rapid polymer relaxation. In Kharbhari et al. [4], it has also been reported that dual-phase diffusion is best suited at different diffusion temperatures for carbon/epoxy mixing. As noted earlier, the basal section shows thermal Fickian diffusion, although at a later stage, a slower increase may be observed. Subsequently, saturation of water absorption can be achieved by synergistic effects of many species of forest organisms, including polymer relaxation, micro-projectors or fillings, and through capillary action

4.2 Factors Affecting Water Absorption

4.2.1 Nanofiller

The nanofiller is a nanoparticle that is added into the composite materials to improve their mechanical properties and reliability. The researchers have used various types of nanofiller in composite material to obtain the desired mechanical property of the composite materials. Nowadays, composite materials are also used in various hydrodynamic environments. In a hydrodynamic environment, the composite materials absorb the water molecules. The absorbed water molecules are present in two forms: free water molecules and bounded water molecules. The absorption of free molecules is due to the presence of voids in the composite materials, these voids help in capillary action to uptake the water molecules, while the bounded water molecules form a hydroxide group with matrix or reinforcement. However, the water absorption property in composite material is affected by the presence of nanofillers, and wt% of nanofillers.

The void content in the composites can affect the moisture absorption behavior of composite materials and can be evaluated by resign burns out test. The void content can be determined using the given Equation 4.26:

$$V_v = 1 - \rho_c \left(\frac{w_f}{\rho_f} + \frac{w_m}{\rho_m} \right),$$

where V_v is the void volume factor present in the composite material, ρ_c is the density of the composite material, ρ_f is the density of fibers, ρ_m is the density of epoxy, w_f is the weight factor of fibers, and w_m weight factor of epoxy.

The amount of water absorbed by the composite is determined by:

$$M_t = \frac{m - m_o}{m_o} \times 100\%,$$

where M_t is the moisture content percentage at time t, m_o is the weight of the specimen at the dry condition, and m is the specimen weight at time t.

The carbon nanotubes (CNTs) and carbon nanofillers (CNFs) are used to fabricate the nano-composite or particle composite. The diameter and length of the CNTs nanofillers are 10 nm and 1 μm, respectively. The diameter of CNF is varied from 20 to 100 nm, and the length is 35 μm. To fabricate the CNTs nano-composite, the CNTs nanofiller is used as a reinforcement. The two different quantities of 0.1 and 0.25 wt% of the CNTs nanofillers are used to fabricate the CNTs nano-composite specimen. Similarly, the 0.25 and 1 wt% of the CNFs nanofillers are used to fabricate the CNFs nano-composite. The epoxy, epoxy/CNTs, and epoxy/CNFs specimens are immersed into the water for 30 days to determine the water uptaking property. It is observed that the moisture absorbed by the epoxy is 1.83%, and the 0.25 and 1.0 wt% epoxy/CNFs are 1.43% and 1.50%, respectively, and the 0.1 and 0.25 wt% epoxy/CNTs are 1.40% and 1.66%, respectively [6].

The TiO₂ nanofiller is embedded into the epoxy of the glass fiber-reinforced polymer composite. The three different amounts 0.1, 0.3, and

0.7 wt% of the TiO_2 nanofiller are added into the glass fiber reinforced polymer (GFRP) composite. As the wt% of the TiO_2 nanofiller increases, the voids contained are also increased, this may happen because the air bubble entrapped between the nanofiller and matrix is unable to come out during the fabrication process. The GFRP with the TiO_2 nanofiller and controlled GFRP composites are submersed into the water of pH 5.65 and maintain a 70°C temperature. The experiment is carried out as per ASTM D 570-98 standard, after a regular time duration, the specimen is taken out from the water, then dried with the help of a dry cotton cloth, and the weight is measured [2].

Figure 4.9. shows the water absorption curve for the controlled glass fiber-reinforced polymer composite and nanofiller TiO_2 GRPF composite (wt%). From Figure 4.9, it is observed that the diffusion coefficient was reduced by 9% with the 0.1 wt% TiO_2 nanofiller. With the addition of 0.1 wt% of TiO_2 in the GFRP composite, it increased a residual flexural strength and interlaminar shear strength by 19% and 18%, respectively.

The water absorption test is carried out on a controlled GFRP composite and a GFRP composite with a Al_2O_3 nanofiller. Both the specimens are immersed into the water for 30 days at a 70°C temperature. With the presence of a 0.1 wt% Al_2O_3 nanofiller into the GFRP composite, the diffusion coefficient is reduced by 10% as compared to the controlled GFRP composite. From the flexural test, it is also determined that the residual flexural and interlaminar shear strength increased by 16% and 17%, respectively [9]. Figure 4.10 shows the water absorption curve by the GFRP composite with and without nanofillers.

Carbon nanotubes are used to improve the mechanical property of composite materials. This nanofiller attracts the attention of many researchers

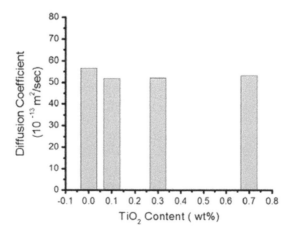

FIGURE 4.9
Water absorption curve for controlled glass fiber-reinforced polymer composite and nanofiller TiO_2 GRPF composite (wt%). (From Nayak, R.K. et al., *Compos. Part A App. Sci. Manuf.*, 90, 736–747, 2016.)

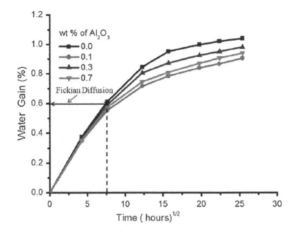

FIGURE 4.10

Water absorption curve by GFRP composite with and without nanofillers. (From Nayak, R.K. and Ray, B.C., *Polymer Bulletin*, 74, 4175–4194, 2017.)

to increase the mechanical behaviors of the composite materials without an increase to its density. Zulfli et al. [8] performed an experiment to determine the water absorption by GFRP and GFRP with a multi-walled carbon nanotube (MWNCT). The immersion experiment is carried out at a different temperature and different % of MWNCT in the GFRP for the duration of 12 months. The water content % in GFRP, GFRP/MWNCT 0.5 wt%, GFRP/MWNCT 1.0 wt%, GFRP/MWNCT 1.5 wt%, and GFRP/MWNCT 2.0 wt% is 1.16%, 1.30%, 1.42%, 1.67%, and 1.54%, respectively. It is observed that as the MWNCT wt% increased, the water absorption is also increased, however, the mechanical property of GFRP with MWNCT increases with the increase in wt%.

Nanosilica is deposited on glass fiber to obtain better bonding between the matrix and the glass reinforcement. Nanosilica acts as a catalyst that produces strong bonds between cyclic butylenes terephthalate resin polymerized cyclic butylenes terephthalate (pCBT) and glass fibers, which increased the mechanical property of the composite materials. At a 60°C temperature in a hydrodynamic environment, the moisture content is 0.491, 1.076, and 0.778 wt% in pCBT, glass fiber (GF)/pCBT, and nanosilica GF/pCBT, respectively. From the comparative study of (pCBT), GF/pCBT, and nanosilica GF/pCBT composites, it is observed that the presence of glass fiber increased the water uptaking property less than the pCBT resin. When the glass fibers are treated with the nanosilica, the water uptaking property decreases [9].

Ashori et al. [10] used 0% to 6% mont morillonite (MMT)-based nanoclay to improve the tensile strength and flexural strength of poly propylene wood floor composite material. They found that the 3% of the nanoclay reduced the water uptaking property, with an increase in tensile strength and flexural strength by 20% and 13%, respectively. The 2 to 5 wt% MMT nanoclay is added with the epoxy/sisal composite material to improve the tensile strength and tensile modulus [11]. With the use of 5 wt% MMT nano-clay, this reduces the water uptake property one third in this composite material and improves the tensile strength and tensile modulus by 27% and 47%, respectively. An unsaturated polyester resin/hemp composite is loaded with 0 to 1.5 wt% of nanoclay to obtain the tensile property. From the study, it is observed that the moisture absorbed by the nanoclay resin/hemp composite reduced by 8%, where the ultimate tensile strength is also decreased by 20% [12]. The MMT nanoclay is used in poly ethylene/bamboo composite material to improve the tensile strength [13]. Ibrahim et al. [14] observed that with the use of 5 wt% MMT nanoclay in poly propylene/sisal composite material, the water absorption reduced and showed better tensile strength. With the use of 3 wt% of nanoclay in a polyester/sisal composite, the mechanical property increased with a reduction in water absorption [15].

With the use of 5 wt% nano-SiO_2 in a high-density poly ethylene/bagasse fiber composite, Hosseini et al. [16] observed that there is such change in the water absorption, where the tensile strength is increased by 71.46%. Kushwaha et al. [17] used a carbon nanotube as a nanofiller in the epoxy/bamboo composite material. The 0.15 wt% of a CNT is added into the epoxy/bamboo composite material, the presence of a CNT nano-material increases the mechanical property of the natural fiber composite in terms of tensile strength, tensile modulus, flexural strength, flexural modulus, and impact strength. The use of a CNT nanofiller in an epoxy/bamboo composite reduced 23.18% of the water absorption with an increase in tensile strength by 6.67%, tensile modulus by 2.7%, flexural strength by 5.8%, flexural modulus by 31%, and impact strength by 84.5%.

4.2.2 Polymer

Polymers are used in the form of a matrix in the composite materials. In composite materials, the matrix is used to hold the fibers together. It gives shape to the composite materials. It also protects the fibers from environmental and mechanical damages. The use of polymer as a matrix gives a lightweight composite, as compared to a metal and ceramic matrix and is also easy to manufacture at less cost. The various types of polymer, such as epoxy [11], vinyl acetate [18], urea formal dehyde [19], polyester [15], etc. are used in composite materials.

Digiycidyl ether of bisphenol As (DGEBAs) are color fever epoxy resins that are used as a matrix in the composite material. This epoxy is immersed into the water for 60 days at a 30°C temperature, and it is observed that the moisture absorbed by the epoxy is 1.83 wt%. With the addition of CNF and CNT nanoparticles into the DGEBA epoxy, its water absorption property reduced [6]. A DGEBA epoxy/glass fiber composite of a 60/40 epoxy glass ratio was immersed into the water for 12 months at a 30°C temperature and absorbed 1.16 wt% water molecules [8]. A 50%/50% DGEBA epoxy/sisal fiber composite was immersed into the water, and it absorbed 62.1% of the water molecules. The addition of nanoclay into the water absorption property of this composite significantly reduced [11].

A low viscosity epoxy resin (FR-251) and epoxy hardener (isophorone-diamine) absorbed 2.34 wt% of a water molecule at room temperature. The water absorption property of the epoxy reduced with the use of nanoclay, HNT, and n-SiC [20]. pCBT resin absorbed 0.147 wt% of the water molecules at a 25°C temperature, and immersion time was 3 months. As the temperature rises to 60°C, the water molecule content is increased to 0.491 wt% [9]. The unsaturated polyester at a 80°C temperature absorbed 3.35 wt% of the water molecules. With unsaturated polyester used in the form of a matrix in a GFRP composite, the water absorption reduced to 2.69 wt% [21].

The water uptaken behavior of a polyamide 6/polypropylene matrix is 7.28% at a 30°C temperature. It is observed that as the temperature rises, the water absorption property reduced. At 60°C, the water absorption is 6.82%, and at 90°C, the water absorption is 6.78% [22]. The water absorption property of a prepolymer tetraglycidyl 4, 4'-diaminodiphenylmethanematrix at different immersion temperatures is determined. From the observation, it is observed that the water absorbed at 20°C, 40°C, 70°C, and 100°C is 2.74, 3.05, 5.32, and 6.21 wt%, respectively [23], which indicates that as the temperature increases, the water absorption of the composite is also increased. The same experiment is carried out for the epoxy resin, water absorption at 19°C, 35°C, 45°C, 60°C, and 90°C temperatures is determined, which is 9.7%, 10.2%, 9.3%, 9.6%, and 8.9% [24].

4.2.3 Temperature

The temperature plays an important role in the moisture absorption of composite materials. The water molecules which are present in the composite materials are in two types: free molecules and bounded molecules. The free molecules of water are trapped into the composite material with the help of capillary action, on the other hand, the bounded molecules form the hydroxide group with the matrix. We know that as the temperature increases, the surface tension of the liquid decreases, which helps to increase the capillary rise. In this section, we discuss the water absorption of different composite materials affected by the temperature change.

The water absorption test is carried out on DGEBA polymer glass fiber composites with and without a MWCNT nanofiller at three different immersed temperatures. Table 4.1 shows the water absorption properties of a GFRP composite at different temperatures and different % of embedded nanofillers. From Table 7, it is clear that as the temperature increases, the water absorption of the composite materials also increases [8].

The pCBT polymer specimens are immersed in the water for 3 months, at the controlled hydrothermal environment. The water absorbed by the polymer at 25°C and 60°C temperatures is 0.147% and 0.491%, respectively, which indicates that the rise in temperature also brings a rise to the water

TABLE 4.1

Temperature Influence on Water Absorption in Composite Material

Materials	Moisture Content %		
	30°C	60°C	90°C
GFRP	1.16	1.64	2.59
GFRP/0.5% MWCNT	1.30	1.81	2.84
GFRP/1.0% MWCNT	1.42	2.15	3.08
GFRP/1.5% MWCNT	1.67	2.43	4.45
GFRP/2.0% MWCNT	1.54	2.74	4.06

absorption property of the epoxy. Similarly, it is observed that at 25°C and 60°C temperatures, the water absorbed by the pCBT/glass fiber composite is 0.856% and 1.076%, respectively. It is observed that in both the conditions, the water absorption increased as the temperature increases [9].

The effect of temperature on the polypropylene polymer and polypropylene organoclay nano-composite at different wt% of organoclay is studied. Table 4.2 shows the water absorption by the polymer and nano-composite at different temperatures is given. From Table 4.2, it is observed that the water absorption by both polymer and nano-composite decreases with an increase in temperature [22].

The tetraglycidyl 4,4′-diaminodiphenylmethane polymer is immersed into the 20°C, 40°C, 70°C, and 100°C hydrothermal environments to determine the water absorption property of the polymer. It is observed that the water absorbed by the polymer at 20°C, 40°C, 70°C, and 100°C temperatures is 2.74%, 3.05%, 5.32%, and 6.21%, respectively. From the above observation, it is concluded that as the temperature increased, the water absorption by the polymer is also increased [23].

TABLE 4.2

Temperature Influence on Water Absorption in Composite Material

Material	Moisture Content % Immersed Temperature		
	30°C	60°C	90°C
PA6/PP	7.28	6.82	6.75
PA6/PP/2wt% nanoclay	10.3	8.53	8.48
PA6/PP/4wt% nanoclay	10.17	8.25	8.10
PA6/PP/6wt% nanoclay	9.84	8.01	7.64
PA6/PP/8wt% nanoclay	8.52	7.57	7.25
PA6/PP/10wt% nanoclay	8.14	7.07	6.85

From the above observation, it is summarized that as the temperature increased, the water absorption is also increased by the composite materials. But in the polyamide 6 (PA6)/polypropylene (PP) polymer, as temperature increased, the water absorption by the composite decreased, thus, this polymer is suitable in a high hydrothermal environment.

4.2.4 Interface

The Field emission scanning electron microscope is used to obtain the interface property of different composite materials. Figure 4.11 shows the interface property of epoxy/glass with and without a Al_2O_3 nanofiller treated in hydrothermal environments.

Figure 4.11a shows the micro cracks/voids and smooth fiber imprints between the fibers and matrix, which lead to decreased interlaminar shear strength and flexural property of GFRP composite materials. With the addition of a nanofiller (AL_2O_3 0.1 and 0.3 wt%), the hackles, fiber breakage, rough fibers imprints, and severely deformed matrix have been observed, which leads to improving the interlaminar shear strength and flexural strength.

Figure 4.12 shows the interface property of GFRP composite materials with or without a TiO_2 nanofiller. The interfacial debonding is present in

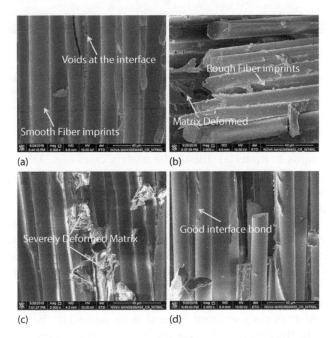

(a) (b)

(c) (d)

FIGURE 4.11
FESEM images of fractured surface of hydrothermally conditioned samples having different wt% of nano-Al_2O_3 (a) -0.0, (b) -0.1, (c) -0.3, and (d) -0.7. (From Nayak, R.K. and Ray, B.C., *Polymer Bulletin*, 74, 4175–4194, 2017.)

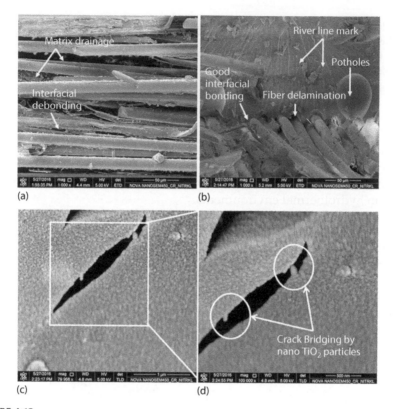

FIGURE 4.12
Interface property of GFRP composite materials without (a) and with a TiO_2 nanofiller (b, c, and d). (From Nayak, R.K. et al., *Compos. Part A App. Sci. Manuf.*, 90, 736–747, 2016.)

the GFRP composite material due to the matrix cracking. As the nanofiller is added into the GFRP composite materials, it creates crack-bridging, which leads to a decrease in crack growth, due to this, the flexure and interlaminar shear strength increased.

Figure 4.13 shows the interfacial property of high density polyethylene/bagasse fiber composite materials. From Figure 4.13, it is observed that the crack and air gap is present at the interface of the matrix and fiber, which leads to an increase in the water absorption property of composite materials.

Figure 4.14 represents the surface roughness of fibers in poly propylene/sisal composite materials without (a) and with nanoclay (b, c, and d). It is observed that in the absence of nanoclay less smoothness occurs, and as the wt% of the nanoclay increased, there was an increase in roughness, which leads to improving the toughness of the composite materials [14].

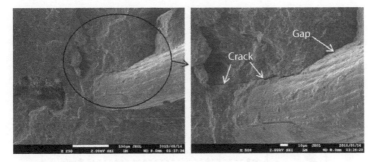

FIGURE 4.13
Interfacial property of high density polyethylene/bagasse fiber composite materials. (From Ibrahim, I.D. et al., *J. Polym. Environ.*, 25, 427–434, 2017.)

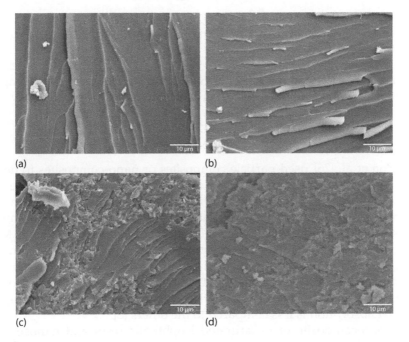

(a)   (b)

(c)   (d)

FIGURE 4.14
Surface roughness of fibers in poly propylene/sisal composite materials without (a) and with a nanoclay (b, c, and d). (From Haq, M. et al., *Compos. Sci. Technol.*, 68, 3344–3351, 2008.)

Graphene nanofiller is used to improve the mechanical property of epoxy/jute composite materials. From the interface study, it is observed that in the absence of the nanofiller, there is no restriction in the crack growth. In nanocomposites, the nano filler makes the bridges that restrict the crack growth,

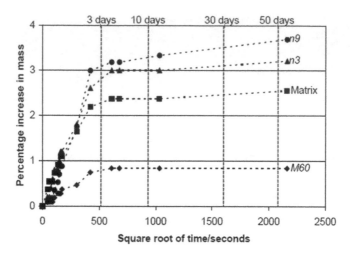

FIGURE 4.15
Water absorption property of a composite in a humid environment. (From Zou, C. et al., *IEEE Trans. Dielectr. Electr. Insul.*, 15, 106–117, 2008.)

and that is the reason in the presence of a nanofiller, the epoxy/jute composite materials show better mechanical property then the controlled epoxy/jute composite [25].

4.2.5 Environment

Chen Zon et al. [26] used particle composites or nano-composites to determine the water absorption property at 75% relative humidity. The equal epoxy and hardener % are taken to form a matrix. SiO_2 was used as a nanofiller or as a reinforcement, the 3 wt% nanofiller in the composite was named n3, the 9 wt% nanofiller was named n9, and the nanofiller of 60 wt% is M60. Figure 4.15 shows the behavior of the epoxy and epoxy with nanofiller in 75% relative humid environment. It is clearly observed that most of the water particles are absorbed in the initial stage, then the water absorption gets slowed down suddenly. It is observed that with the presence of a 60 wt% nanofiller, the water absorption is less compared to the others [8].

The nano-composite of unsaturated isophthalic resin and nanosilica is used to determine the water absorbing property in salt water and compare it with a distilled water environment. The 1, 2, and 3 wt% of SiO_2 is used to fabricate a nano-composite. It is observed that the nano-composite with 3 wt% of SiO_2 absorbed a less amount of water as compared to the others. When the composite is in contact with the distilled water environment, the water absorption is higher as compared to the salt water. It is observed that as the salinity % increased, the water absorption of the composite material decreased, this is because the salt molecules restrict water molecules to rice in the void or crack present in composite materials [27].

References

1. Shen, C.-H., and G. S. Springer. Moisture absorption and desorption of composite materials. *Journal of Composite Materials* 10.1 (1976): 2–20.
2. Nayak, R. K., K. K. Mahato, and B. C. Ray. Water absorption behavior, mechanical and thermal properties of nano TiO_2 enhanced glass fiber reinforced polymer composites. *Composites Part A: Applied Science and Manufacturing* 90 (2016): 736–747.
3. Grace, L. R., and M. C. Altan. Characterization of anisotropic moisture absorption in polymeric composites using hindered diffusion model. *Composites Part A: Applied Science and Manufacturing* 43.8 (2012): 1187–1196.
4. Tang, W. C., R. V. Balendran, A. Nadeem, and H. Y. Leung. Flexural strengthening of reinforced lightweight polystyrene aggregate concrete beams with near-surface mounted GFRP bars. *Building and Environment* 41.10 (2006): 1381–1393.
5. Jiang, X., H. Kolstein, and F. S. K. Bijlaard. Moisture diffusion in glass–fiber-reinforced polymer composite bridge under hot/wet environment. *Composites Part B: Engineering* 45.1 (2013): 407–416.
6. Prolongo, S. G., M. R. Gude, and A. Urena. Water uptake of epoxy composites reinforced with carbon nanofillers. *Composites Part A: Applied Science and Manufacturing* 43.12 (2012): 2169–2175.
7. Nayak, R. K., and B. C. Ray. Water absorption, residual mechanical and thermal properties of hydrothermally conditioned nano-Al_2O_3 enhanced glass fiber reinforced polymer composites. *Polymer Bulletin* 74.10 (2017): 4175–4194.
8. Zulfli, N. H. M., A. Abu Bakar, and W. S. Chow. Mechanical and water absorption behaviors of carbon nanotube reinforced epoxy/glass fiber laminates. *Journal of Reinforced Plastics and Composites* 32.22 (2013): 1715–1721.
9. Yang, B., J. Zhang, L. Zhou, M. Lu, W. Liang, and Z. Wang. Effect of fiber surface modification on water absorption and hydrothermal aging behaviors of GF/pCBT composites. *Composites Part B: Engineering* 82 (2015): 84–91.
10. Ashori, A., and A. Nourbakhsh. Preparation and characterization of polypropylene/wood flour/nanoclay composites. *European Journal of Wood and Wood Products* 69.4 (2011): 663–666.
11. Mohan, T. P., and K. Kanny. Water barrier properties of nanoclay filled sisal fibre reinforced epoxy composites. *Composites Part A: Applied Science and Manufacturing* 42.4 (2011): 385–393.
12. Haq, M., R. Burgueño, A. K. Mohanty, and M. Misra. Hybrid bio-based composites from blends of unsaturated polyester and soybean oil reinforced with nanoclay and natural fibers. *Composites Science and Technology* 68.15–16 (2008): 3344–3351.
13. Han, G., Y. Lei, Q. Wu, Y. Kojima, and S. Suzuki. Bamboo–fiber filled high density polyethylene composites: Effect of coupling treatment and nanoclay. *Journal of Polymers and the Environment* 16.2 (2008): 123–130.
14. Ibrahim, I. D., T. Jamiru, R. E. Sadiku, W. K. Kupolati, and S. C. Agwuncha. Dependency of the mechanical properties of sisal fiber reinforced recycled polypropylene composites on fiber surface treatment, fiber content and nanoclay. *Journal of Polymers and the Environment* 25.2 (2017): 427–434.

15. Venkatram, B., C. Kailasanathan, P. Seenikannan, and S. Paramasamy. Study on the evaluation of mechanical and thermal properties of natural sisal fiber/general polymer composites reinforced with nanoclay. *International Journal of Polymer Analysis and Characterization* 21.7 (2016): 647–656.
16. Hosseini, S. B., S. Hedjazi, L. Jamalirad, and A. Sukhtesaraie. Effect of nano-SiO_2 on physical and mechanical properties of fiber reinforced composites (FRCs). *Journal of the Indian Academy of Wood Science* 11.2 (2014): 116–121.
17. Kushwaha, P. K., C. N. Pandey, and R. Kumar. Study on the effect of carbon nanotubes on plastic composite reinforced with natural fiber. *Journal of the Indian Academy of Wood Science* 11.1 (2014): 82–86.
18. Vilakati, G. D., A. K. Mishra, S. B. Mishra, B. B. Mamba, and J. M. Thwala. Influence of TiO_2-modification on the mechanical and thermal properties of sugarcane bagasse–EVA composites. *Journal of Inorganic and Organometallic Polymers and Materials* 20.4 (2010): 802–808.
19. Jiang, Y., G. Wu, H. Chen, S. Song, and J. Pu. Preparation of nano-SiO_2 modified ureaformaldehyde performed polymer to enhance wood properties. *Reviews on Advance Materials Science* 33.1 (2013): 46–50.
20. Alamri, H., and I. M. Low. Effect of water absorption on the mechanical properties of nano-filler reinforced epoxy nanocomposites. *Materials & Design* 42 (2012): 214–222.
21. Faguaga, E., C. J. Pérez, N. Villarreal, E. S. Rodriguez, and V. Alvarez. Effect of water absorption on the dynamic mechanical properties of composites used for windmill blades. *Materials & Design (1980–2015)* 36 (2012): 609–616.
22. Chow, W. S., A. Abu Bakar, and Z. A. Mohd Ishak. Water absorption and hygrothermal aging study on organomontmorillonite reinforced polyamide 6/polypropylene nanocomposites. *Journal of Applied Polymer Science* 98.2 (2005): 780–790.
23. Nogueira, P., C. Ramirez, A. Torres, M. J. Abad, J. Cano, J. Lopez, I. López-Bueno, and L. Barral. Effect of water sorption on the structure and mechanical properties of an epoxy resin system. *Journal of Applied Polymer Science* 80.1 (2001): 71–80.
24. Popineau, S., C. Rondeau-Mouro, C. Sulpice-Gaillet, and M. E. Shanahan. Free/bound water absorption in an epoxy adhesive. *Polymer* 46.24 (2005): 10733–10740.
25. Shen, X., J. Jia, C. Chen, Y. Li, and J. K. Kim. Enhancement of mechanical properties of natural fiber composites via carbon nanotube addition. *Journal of Materials Science* 49.8 (2014): 3225–3233.
26. Zou, C., J. C. Fothergill, and S. W. Rowe. The effect of water absorption on the dielectric properties of epoxy nanocomposites. *IEEE Transactions on Dielectrics and Electrical Insulation* 15.1 (2008): 106–117.
27. Maheshwari, N., S. Neogi, M. P. Kumar, and D. Niyogi. Study of the effect of silica nanofillers on the sea-water diffusion barrier property of unsaturated polyester composites. *Journal of Reinforced Plastics and Composites* 32.13 (2013): 998–1002.
28. Kumosa, L., B. Benedikt, D. Armentrout, and M. Kumosa. Moisture absorption properties of unidirectional glass/polymer composites used in composite (non-ceramic) insulators. *Composites Part A: Applied Science and Manufacturing* 35.9 (2004): 1049–1063.

5

Hydrothermal Effect on Mechanical Properties of Organic Nanofillers Embedded FRP Composites

5.1 Introduction

It is well established that zero-dimensional, one-dimensional, i.e., nanotubes and nanowires, and two-dimensional, i.e., nanosheets, have been developed to expand the several properties of composite fibers, especially carbonaceous-based nanofillers such as carbon nanotubes (CNTs), graphene, graphite [1–4]. CNTs have shown as effective nanofillers that improve the matrix-dominated properties in the fibrous composites in terms of their electrical and mechanical behavior. Due to the outstanding conjugated hollow structure, better conductivity and high specific surface area make CNT-based nanofillers as one of the potential fillers to be used as property enhancement. As the manufacturing cost of producing high-grade CNTs in bulk amounts is relatively high, this causes restrictions in the use of these nanofillers. This high cost could be minimized by a newly evolved allotrope of carbon named grapheme, and it can be one of the solutions to this cost constraint. Graphene is a two-dimensional, single-atom layered, and highly conjugated carbon nanomaterial, whose exploration history can be outlined back to 1859. The excellent properties of nanoparticles and nano-scale dimensions have helped in designing innovative methods and continuous improvements in existing techniques. Several issues related to the in-service environment are challenging tasks to manage. Environmental challenges evolved as the main problem and these could have solutions from novel nanomaterials in numerous areas such as energy production, water treatment, and contaminant sensing [5].

Moisture uptake from the environment through direct contact or indirect contact is one of the detrimental effects of fiber reinforced polymer (FRP) polymeric composites. Basically, water acts as a plasticizer for the polymeric matrix, reducing its reliability and stability. Numerous methods and models have been developed for water absorption in polymers and their composites [6–11]. At the preliminary phase of moisture ingression, glass transition temperature and other mechanical properties (strength, modulus) start to damage due to plasticization [12]. FRP composites can absorb moisture through the matrix and may follow Fickian diffusion kinetics [13]. The existence of cracks and voids can speed up the diffusion process and stimulates the non-Fickian diffusion kinetics [14,15]. Several authors have also stated the wicking of moisture through the fiber/matrix interface to support the non-Fickian diffusion. The moisture along the fiber/matrix interface and between adjacent plies, exploits microcracks, voids, and the capillaries and results in the weakening of these composites [16]. These water molecules are characterized as type I bound water. The type I bound water molecules interact with the epoxy resin by van der Waals force and hydrogen bonding, and thus enhances the mobility of polymeric chain segments. Once the polymer becomes moisture saturated, a recovery in mechanical properties takes place due to the connection of the water molecule with various hydrophilic groups of the polymer (such as amines and hydroxyls). These water molecules having multiple chemical connections are termed as type II bound water molecules. These multiple connections reduce the flexibility of the polymeric chains, and hence improve the mechanical properties. In the polymeric material, the degree of moisture diffusion can be calculated by the holes for the accommodation or active sites of water molecules in the cross-linked network structure of the polymer [17]. In the amorphous or semi-crystalline polymer, it may be possible during curing that the polymeric chains are not closely packed and leave some holes, nanopores, or free volume with them.

The objective of this chapter is to discuss the hydrothermal behavior on mechanical properties of organic nanofiller [graphene, graphite, single-walled CNT (SWCNT), multi-walled CNT (MWCNT)]-embedded FRP composites. This study explores the water uptake kinetics of epoxy composites reinforced with graphene, graphite carbon nanofillers, and also with the combination of different organic nanofillers. The effects of water ingress on the mechanical behavior of these materials have been explored. Different organic nanofillers such as graphene, graphite, SWCNTs, MWCNTs, and a combination of various nanofillers have been used in order to determine the influence of the nanofillers nature and their morphology.

5.2 Graphene

The word graphene was created as an arrangement of the word graphite and the suffix-ene by Hanns-Peter Boehm, who defined single-layer carbon foils in 1962 [18,19]. The attention toward the use of graphene increased after the excellent electronic properties of monolayer graphite sheets in 2004 were reported by Novoselov and Geim [20–22]. The exceptional plane geometry and structure of monolayer graphene reveal its excellent properties such as high Young's modulus (1100 GPa), high fracture strength (125 GPa), excellent thermal (5000 W/mK) and electrical conductivity (10^6 S/cm), large specific surface area (theoretically calculated value, 2630 m^2/g), and fast mobility of charge carriers (200,000 cm^2/Vs) [23,24]. The outstanding aforesaid behavior of graphene creates a wide range of applicability in various sectors like energy storage and conversion, electronics, biotechnology, membranes, flexible wearable sensors, actuators, and exclusively in the development of fiber polymeric-based composite materials. By reinforcing graphene into the polymeric composites, it became possible to achieve multifunctional properties in composites. Generally, the following three parameters decide the behavior of graphene-based polymer nano-composites:

- Nanoscopic detention of polymer matrix chains;
- Nano-scale inorganic constituents and deviation in properties; as stated by several investigations of their substantial variation relating to their size; and
- Nanoparticle arrangement and establishment of large polymer/ particle interfacial area.

The preparation of graphene is carried out through different techniques, out of which the frequently used methods are the reduction of graphene oxide (GO), exfoliation of graphite, and chemical vapor deposition. But, the lower dispersibility of graphene in organic and inorganic solvents creates a bottleneck, and that has to be sorted out before its usage in a range of applications. In some solvents, a better dispersibility of graphene led to an appropriate and homogeneous development of better nano-composites. Therefore, it is of utmost importance to alter the graphene solubility in different solvents to execute good dispersibility, which will be critical and crucial for commercial applications.

The modification in graphene is usually carried out with the help of covalent and non-covalent techniques. The non-covalent technique contains p–p stacking interfaces, hydrogen bonding, coordination bonds, electrostatic interaction, and van der Waals force. This variation technique can highly reserve the graphene's natural structure; meanwhile, the several interactions among different functional groups and usually graphene surfaces are reasonably weaker. Since the surface of the graphene is weak; hence, it is not appropriate for highly critical applications where strong bond interactions are mandatory. Covalent techniques can be utilized to generate different composites with strong bond interactions between the graphene surface and the modifier. However, graphene's unique arrangement is generally destructed, leading to compromised electrical conductivity and mechanical properties.

Usually, the natural polymeric fibers contain cotton, silk, linen, and hair, whereas the synthetic fibers include polyamide, polyester, polyacrylonitrile, polypropylene, and polyethylene and are extremely essential in the present decade and widely used in the field of textiles, medical, packaging industries, and structural applications. The broad application of these polymeric fibers not only depends upon their low density and high strength, but also on abrasion resistance, durability, chemical, and environmental stability. In the case of conductive, thermostable properties and toughness, this is required for critical applications. Embedding various nanofillers like CNTs and graphene into the polymeric composites certainly improves the multifunctional properties in terms of mechanical, electrical, storage, sensing, and actuation [25–27]. Improved fiber toughness, i.e., ability to withstand mechanical energy before rupture will indeed be used for different ranges of high critical applications. Hence, graphene is one of the alternatives to be used with fiber polymeric-based materials, and it has been investigated by various researchers [28,29]. The composite fibers are primarily prepared with two procedures; spinning and mixing, as revealed in Figure 5.1. Graphene

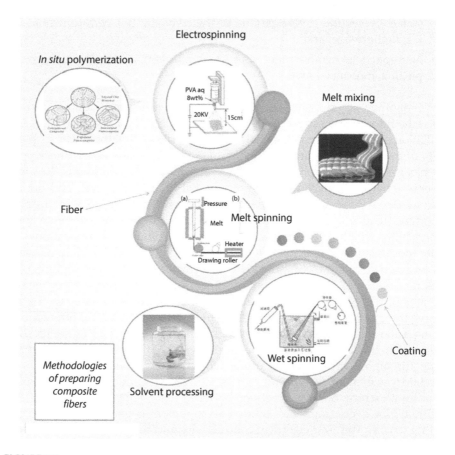

FIGURE 5.1
Schematic illustration of the methods for the preparation of fiber composite materials. (From Ji, X. et al., *Compos. Part A Appl. Sci. Manuf.*, 87, 29–45, 2016.)

can be further mixed at the time of the mixing process in the following techniques; in situ polymerization, solvent mixing, and melt processing. Few have tried the coating of graphene in the fiber surface by the spinning methodology. The spinning methods may be broadly classified as melt-spinning, wet-spinning, and electrical-spinning.

On the incorporation of CNTs [30], graphene showed the better among the two in terms of the fiber's mechanical behavior owing to the graphene's unique structure [31]. The mechanical properties of graphene were evaluated using atomic force microscopy (AFM) and recorded a breaking strength of 130 GPa and Young's modulus of 1.1 TPa [32]. The elastic modulus of chemically reduced monolayers of graphene was calculated as 0.25 TPa [33]. Figure 5.2 illustrates the basic structure of all carbon materials that can be rolled up into: (a) zero-dimensional fullerenes, (b) one-dimensional CNT, and stacked into (c) three-dimensional graphite.

Several investigators have revealed various outcomes on the mechanical behavior of graphene/polymer fiber composites. A number of techniques that include the graphene interaction with fibers by in situ polymerization,

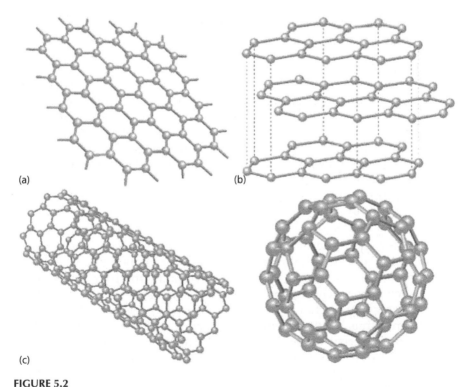

(a)   (b)

(c)

FIGURE 5.2
Graphene: The basic structure of all carbon materials that can be rolled up into (a) zero-dimensional fullerenes, (b) one-dimensional CNT, and stacked into (c) three-dimensional graphite. (From Geim, A. and Novoselov, K., *Nat. Mater.*, 6, 183–191, 2007.)

melt mixing, solvent mixing, and coating are important for fiber-dominated properties in terms of mechanical properties. The coating of graphene on the fiber surface usually occurs by spraying and the "dip and dry" technique. The addition of raw graphene or functionalized graphene to the raw fiber resulted in graphene/fiber composites that showed enhanced mechanical behavior. The chances of environmental decontamination were higher when graphene-based materials were used with sorbent or photocatalytic materials. The very high electron mobility (10,000–50,000 cm^2/Vs) of graphene particles with supportable current densities makes graphene able to be used in a vast range of applications with respect to the environment. These graphene-based materials are used as adsorbents to eliminate organic compounds, metal ions, and harmful gases such as carbon dioxide mostly from the water environment. In the investigations related to the blockage of liquid and gas, graphene nano-composites are useful with resistant proof. Ultra-thin graphene established water separation membranes like stacked graphene oxide membranes and a non-porous graphene membrane was developed where electrostatic interactions and size exclusion composition comprise the main criteria in identifying the best one. The reaction of the graphene-based nanomaterial with lipid membranes of microbes size significantly affects the anti-microbial activity, as per studies [5].

FRP composites are being extensively used in the aircraft and aerospace industries. In the case of low-temperature applications, especially in space vehicles, substituting metallic cryogenic fuel tanks with FRP materials for reusable space launch vehicles and expendable space launch vehicles is the technique to create lightweight, high-strength, and durable structures. For example, the temperature range of a satellite lies between −160°C and +130°C. This occurs as the satellites cross the threshold of the earth's shadow, by which high amplitude thermal cycles become durable. Carbon fiber reinforced polymer (CFRP) composites possess higher ratios of strength-to-weight and stiffness to weight. Finally, these FRPs-based materials find an extensive usage signifying thermal cycles with amplitudes exceeding 300°C. Thermosetting polymers are usually used as a matrix for CFRP composites since they are more thermally stable, chemically inert, and have moderate mechanical and electrical properties. The incorporation of a very small amount of nanofillers into the matrix phase certainly improves the properties of the composite. During low-temperature atmospheres, thermosetting polymers show brittle behavior because of their rigid cross-linked molecular arrangements, thus making them easily cracked subject to thermal-fatigue loading [34]. In recent times, graphene attracted a lot of attention owing to their excellent electrical, mechanical, and thermal properties. The large surface area of graphene-based nano-composites as compared to other nanofillers makes it a proper nanoparticle as reinforcement for polymeric composites [35,36].

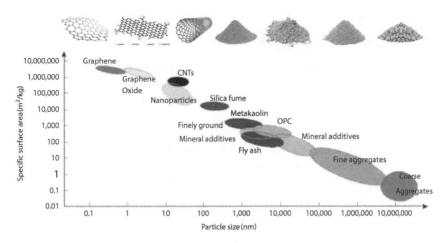

FIGURE 5.3
Reveals the specific surface area of various particles used for engineering applications. (From Mohan, V.B. et al., *Compos Part B*, 142, 200–220, 2018.)

It was observed from Figure 5.3 that graphene and GO possess the smallest particles with relatively larger surface area as compared to other particles. GO is basically the oxidized form of graphene that is used as filler material for polymeric-based composites. It is made up of oxidative fragments of functionalized graphene sheets. GO possesses large surface area (2630 m^2/g), high Young's modulus (~1.0 TPa), high intrinsic mobility (200,000 cm^2/V), high optical transmittance (~97.7%), and thermal conductivity (~5000 W/mK) [37]. The preparation of GO is generally carried out by the treatment of graphite flakes with potassium permanganate sodium nitrate and anhydrous sulfuric acid where the polar groups are introduced onto the graphite's surface, thereby widening the interlayer spacing of the graphene planes [38]. GO exhibits excellent handling characteristics and high chemical reactivity due to the existence of carboxyl, epoxide, hydroxyl, and carbonyl functional groups [39,40]. The chemical coating of GO on the surface of a carbon fabric or carbon fiber certainly improves the interfacial properties between the fiber/polymer interface [41]. Hung et al. [42] fabricated GO with a coating on the fabrics surfaces by an electrophoretic deposition technique, to increase the interlaminar shear strength of the composites. They reported that at a low-temperature environment, the ultimate tensile strength and Young's modulus of the composites were found to be increased, which is the in-service temperature range of typical aircraft at a tropopause region. Wu et al. [34] have investigated that using graphene nanoplatelets (GNPs) certainly improves the mechanical properties and the critical stress intensity factor (K_{IC}) of Al$_2$O$_3$ in a low-temperature environment. A GNP, here, was acting as a bridging nanofiller to connect up crack surfaces together to resist any crack propagation in Al$_2$O$_3$.

5.3 Graphite

Graphite or carbon fibers are high-modulus materials that are electrically and thermally conductive. They can withstand flexing within their elastic limits very well [43]. Graphite fibers are one of the most extensive fiber forms in the applications of high-performance composite structures. Graphite fibers can be manufactured with a varied range of properties; however, they usually reveal higher compressive and tensile strength, high moduli, excellent fatigue resistance, and do not corrode [44]. Graphite fibers are made up of pitch, rayon, or polyacrylonitrile precursors. As the polyacrylonitrile fibers are more expensive as compared to rayon fibers, they are widely used for structural carbon fibers because their carbon yield is almost twice that of rayon fibers. The pitching method yields fibers that have a lower strength than those created from polyacrylonitrile, but it can generate ultra-high-modulus fibers (50–145 msi).

Graphite fibers have a carbon percentage above 99% and show high elastic moduli, while carbon fibers have carbon contents between 93% and 95% [45]. The key advantages of carbon fibers are high-modulus-weight ratios, high-strength-weight ratios, excellent corrosion resistance, and very good fatigue strength. But the cost of these fibers is affected by the high price of raw materials and the lengthy methods of graphitization and carbonization. Furthermore, graphite fibers cannot be simply wetted by the matrix, hence, sizing is essential for formally embedding them in the matrix. Generally, carbon fibers are comprised of amorphous carbon and graphitic carbon; their high modulus is given by the graphitic form in which carbon atoms are organized in a systematic crystallographic structure of parallel layers, as shown in Figure 5.4.

Natural graphite flakes (NGFs) are present a great deal in nature, they are easy-to-process two-dimensional materials with high-strength,

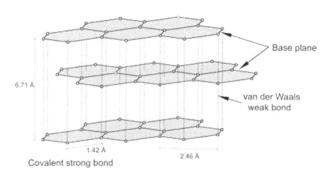

FIGURE 5.4
Crystal structure of graphite. (https://www.sciencedirect.com/topics/engineering/graphite-fiber)

low cost, high-modulus values, low thermal expansion coefficients, low deformability at high temperatures, and high aspect ratios [46–50]. Thus, NGFs possess high fracture toughness, high thermal conductivity, and excellent thermal shock resistance. Graphite involves both a prismatic and basal plane, a prismatic has higher activity as compared to a basal plane. It suggests that functional groups can be certainly generated in an edge site. Therefore, it can be said that natural graphite has more hydrophilic properties than artificial ones, which are detachable functional groups through heat treatment at around 2800°C. In this nature, NGFs have benefits to better the interfacial adhesion with a polymer matrix on fabricating composites. The uniform distribution of NGFs in an epoxy matrix substantially improves the interfacial mechanical properties of FRP composites. Thus, NGFs are regarded as excellent materials for high-temperature applications [51–54].

Seong Hwang Kim et al. [55] studied the effect of hydrophilic graphite flakes on the thermal conductivity and fracture toughness of basalt fibers (BFs)/epoxy composites. They have reported that NGFs certainly increased the mechanical properties of the composites, i.e., the critical strain energy release rate (G_{IC}) and critical stress intensity factor (K_{IC}). Specifically, basalt FRP composites with 20 wt% of natural graphite flakes showed the highest mechanical behavior among all other composites. The fracture toughness mechanisms were showed by a crack theory based on the various morphological analyses of the fracture surfaces. Furthermore, the graphite enhanced composites revealed high thermal conductivity and thermal stability. This may be attributed to the outstanding thermal conductivity and stability of the graphite flakes.

Figure 5.5a and b illustrate the effect of K_{IC} and G_{IC} on the fracture toughness of NGF/BFs/epoxy composites with various NGF contents. Figure 5.5a indicates that the strength of BFs/epoxy composites can be increased further by incorporating NGFs. The NGB-20 composites showed the highest value of K_{IC} 90.56 MPam1/2, 83% higher than the

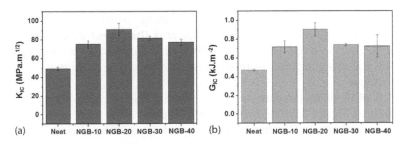

FIGURE 5.5
Mechanical properties of NGF/BFs/epoxy composites as a function of NGF content: (a) K_{IC} and (b) G_{IC}. (From Kim, S.H. et al., *Compos. Part B Eng.*, 153, 9–16, 2018.)

49.44 MPam1/2 K_{IC} value of the neat sample. As revealed in Figure 5.5b, the NGB-20 composites had the highest value of 0.9 kJm^{-2} G_{IC}, which is 92% higher than the neat sample 0.47 kJm^{-2}. Figure 5.5a and b depict similar results for K_{IC} and G_{IC}. Furthermore, more loading increases in the NGFs content than the NGB-20 composites reduce the fracture toughness of the composites. As the NGF loading increases, toughness is reduced because of overloading beyond the maximum fracture toughness. In addition, the distance between the NGF layers results in a cohesive fracture due to a combination of delamination, crack deflection, and shorter inter-particle distances [56–60].

The fracture surface of the sample after the K_{IC} test was detected by scanning electron microscopy images (SEM), as shown in Figure 5.6. Figure 5.6a shows the fracture of the control specimen, and Figure 5.6b illustrates a magnified image of the enclosed area of Figure 5.6a. The cross-section of the control specimen reveals a smooth surface, and brittle in nature, as observed owing to regular crack growth. Figure 5.6c represent the SEM image of NGB-20 composites and Figure 5.6d revealed the magnified image of the enclosed area as shown in Figure 5.6c. The scattering of irregular and side-branch cracks was confirmed and attributed to the requirement of high energy at the time of propagation of the crack. This reveals that there is no agglomeration of epoxy interfacial interactions and NGFs and the dispersion was uniform, thus confirming the better load transfer of BFs. Figure 5.6e shows the fracture of the NGB-40 composites, and Figure 5.6f shows magnified images of the enclosed region in Figure 5.6e. As shown in Figure 5.6e and f, the excessive amount of NGFs in the epoxy matrix agglomerated, which made it tough for them to have be well dispersed, thus lowering the mechanical interfacial properties.

Zhou et al. [6] studied the effects of a water environment on moisture (H_2O) absorption characteristics of a unidirectional T300/934 graphite/epoxy composite. Water sorption in graphite/epoxy (T300/934) material showed both Fickian and non-Fickian diffusion behavior. Moisture-induced expansion of a T300/934 composite was measured in length (fiber direction), thickness, and width directions. Basically, no expansion due to water absorption was detected in the fiber direction dimension. Substantial dimensional variations resulting from moisture-induced expansion were perceived in the width and thickness directions of the laminate. A crack/mass-loss model positively defines the phenomenological behavior of a graphite/epoxy composite resulting from water absorption processes.

Aspects of non-Fickian diffusion behavior for the graphite/epoxy material are shown in Figure 5.7. The solid lines in Figure 5.7 represent the theoretical Fickian diffusion. The symbols represent experimental data. In stage I,

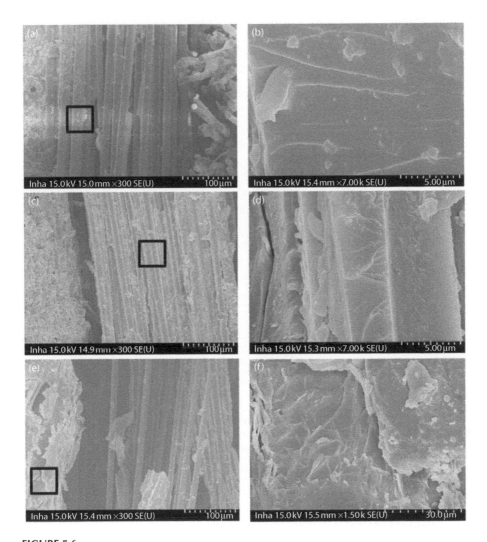

FIGURE 5.6
SEM images of fracture surfaces: (a) neat sample (c) NGB-20 composite (e) NGB-40 composite. (b), (d), (f) are magnified images of enclosed regions in (a), (c), (e), respectively. (From Kim, S.H. et al., *Compos. Part B Eng.*, 153, 9–16, 2018.)

moisture diffusion is characteristic of Fickian behavior. While in stage II, non-Fickian behavior was related with microcrack formation, particularly at the free edge of the tested material. In stage III, the non-Fickian behavior was categorized by the gradual weight loss of the sample resulting from resin leaching and dissolution.

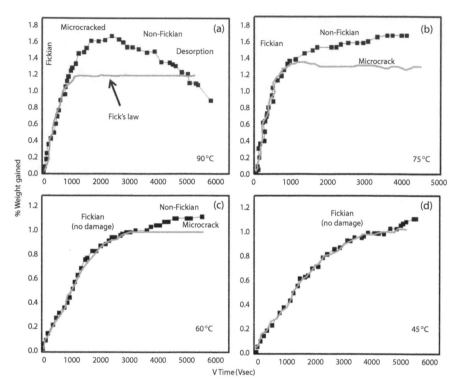

FIGURE 5.7
Moisture absorption profiles of a graphite/epoxy laminate immersed in distilled water at (a) 45, (b) 60, (c) 75, and (d) 90°C for more than 8000 h, showing experimental data and a theoretical prediction of Fickian diffusion. (From Zhou, J. and Lucas, J.P., *Compos. Sci. Technol.*, 53, 57–64, 1995.)

5.4 Single-Walled Carbon Nanotubes

The discovery of carbon nanotubes showed that the new arrangement of carbon atoms can encounter other traditional materials used for a wide range of applications. The exceptional thermal, electrical, optical, and mechanical properties of CNTs have garnered substantial attention, and this motivated the researchers and scientists to excel the use of these materials in day-to-day life. But, CNT aggregates are composed of an extensive sort of CNT types, which has a vigorously negative impact on the observed performance of macroscale devices made from them. Therefore, it has become important that various types of CNTs can be sorted in accordance with their diameter, length, chiral angle electrical behavior, and even-handedness, which will gain interest in them [61].

SWCNTs are quasi-one-dimensional materials, which constitute one of the allotropic forms of carbon. They can be perceived as seamlessly rolled up

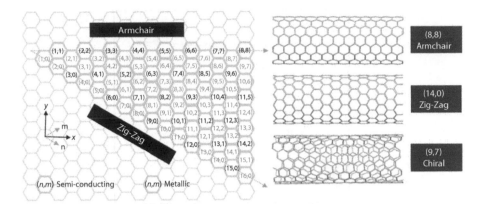

FIGURE 5.8
Helicity map of SWCNTs of different (n, m) chiral vectors. (From Janas, D., *Mater. Chem. Front.*, 2, 36–63, 2017.)

hollow cylinders made of graphene (a sheet of carbon atoms) with diameters ranging between 0.4 and 5 nm [62]. These nanostructures have shown outstanding electrical [63–66], thermal [67–70], optical [71–73], and mechanical [74–76] properties, and have driven strong research and investigation all around the globe.

Usually, CNTs originate in an extensive range of varieties, subjecting how they are accumulated from individual carbon atoms. The easiest method to categorize them is based on the way the intangible arrangement of the graphene sheet is rolled up. Which is called chiral vector or chirality, which is denoted by (n, m) indices, describes the number of unit vectors along two directions of the hexagonal lattice to form a CNT as shown in Figure 5.8.

5.5 Multi-Walled Carbon Nanotubes

Prolongo et al. [12] reported on water absorption monitored gravimetrically and its effect on the thermal and mechanical properties. They found that the maximum water content absorbed significantly decreases with the addition of nanofillers, especially in epoxy/CNT composites. This decrease was revealed both in terms of moisture diffusivity in the material and the saturated moisture content, as shown in Figure 5.9. In the case of a CNT/epoxy system, the resistance to moisture absorption may be attributed to several factors. The presence of hydrophobic CNTs in the epoxy matrix at numerous positions delivers a very uncomfortable path for the water molecules to diffuse inside the material. The level of this blockage property is positively

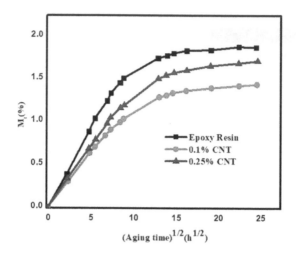

FIGURE 5.9
Water uptake of epoxy resin and nano-composites reinforced with CNTs. (From Prolongo, S.G. et al., *Compos. Part A Appl. Sci. Manuf.*, 43, 2169–2175, 2012.)

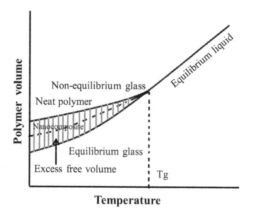

FIGURE 5.10
Volume–temperature behavior of an amorphous neat polymer and nano-composite. (From Starkova, O. et al., *Eur. Polym. J.*, 49, 2138–2148, 2013.)

a function of the specific surface area of the CNTs; the higher the specific surface area, the better is the hindrance to the water molecules. Furthermore, the influence toward the higher moisture resistance of a CNT/epoxy composite comes from the reduction in the free volume [12] of the epoxy due to the addition of CNTs, which occupy those free spots (explained in Figure 5.10). These free spots in the polymer are probable places for water uptake. Uniform dispersion of CNTs efficiently reduces these free sites and later delays the moisture absorption propensity.

Zulfli et al. [77] studied the mechanical and water absorption behaviors of a CNT-reinforced epoxy/glass fiber composite. The epoxy laminates were fabricated by the hybridization of glass fibers and MWCNTs. The flexural behavior, fracture toughness, and impact strength of the epoxy/glass fiber composites were improved significantly by the insertion of MWCNTs. The composite specimens were immersed in a water bath at different temperatures (30°C, 60°C, and 90°C).

The water absorption kinetics of the MWCNT-reinforced epoxy/glass fiber composites followed the Fickian law behavior. The incorporation of MWCNTs enhances the equilibrium moisture content (M_m) and diffusion coefficient (D), but lowers the activation energy of water diffusion (E_a) for epoxy/glass fiber composites.

Starkova et al. [78] have explored the research on the effect of MWCNTs on the water transport behavior of epoxy resin at different relative humidities and temperatures. They have reported that the existence of MWCNTs decreases the moisture diffusivity. But at the saturation point, the moisture content for both is found to be almost equal in a neat epoxy resin and a MWCNT-modified epoxy resin, as revealed in Figure 5.11. They have also confirmed this performance of a MWCNT-enhanced epoxy resin at different temperatures and relative humidities. This water transport phenomenon of a MWCNT/epoxy nano-composite is described by the "free volume" and epoxy-water interactions (Figure 5.10b).

Prusty et al. [79] studied the water-induced damage and degradations in MWCNT-filled glass FRP composites. They have investigated the effect of diffusion temperature on the water uptake kinetics and the succeeding degradation behavior of a MWCNT-enhanced glass FRP (MWCNT-GFRP) composite. The existence of MWCNTs in the GFRP composite significantly

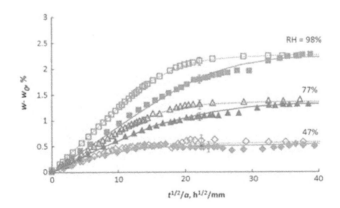

FIGURE 5.11
Moisture uptake by the neat epoxy (open symbols) and nano-composite filled with 0.5 wt% of MWCNTs (filled symbols) under different relative humidity, $T = 20$°C. (From Starkova, O. et al., *Eur. Polym. J.*, 49, 2138–2148, 2013.)

obstructed its water absorption propensity at lower aging temperature (25°C). However, MWCNT reinforcement in a GFRP composite unfavorably affected its high-temperature water resistance due to the generation of adverse hygroscopic and thermal stresses at the MWCNT/polymer interfaces.

The water uptake kinetics of both GFRP and MWCNT-GFRP composites at 25°C and 90°C temperatures are shown in Figure 5.12. Low-temperature conditioning (LTC) shows positive effects of MWCNT incorporation in a GFRP composite both in terms of water uptake rate and water saturation content. The decreased water affinity of the MWCNT-enhanced GFRP composite comes from the intrinsic resistance obtained by the hydrophobic nature of MWCNTs toward water ingression. CNTs possess a very high aspect ratio, and these nanotubes exert forces to the water molecules to follow a tortuous extended path [12] for its further penetration into the bulk material, which in turn is revealed in a lesser water diffusivity rate of the MWCNT-GFRP than that of the control GFRP composite at this temperature. This suppression in water absorption kinetics of the GFRP composite owing to a MWCNT filler may also be clarified from the "free volume"

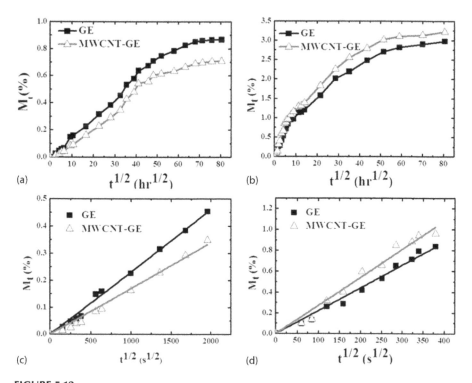

FIGURE 5.12
Water uptake kinetics of GE and MWCNT-GE samples at (a) 25°C and (b) 90°C. Linear fitting of the initial kinetic data at (c) 25°C and (d) 90°C. (From Prusty, R.K. et al., *J. Appl. Polym. Sci.*, 135, 45987, 2018.)

theory of polymer. The equilibrium water content and rate of diffusion are maintained by the dispersion and total volume of free sites in the polymeric chain network at the specified temperature range. Polymeric materials embedded with nanofillers have a tendency to attend its thermodynamic equilibrium stage more closely than the bulk polymer. This, in turn, reduces the additional free volume of the polymer and therefore enhances its water resistivity [78,80].

Aging to different temperatures is expected to have an impact on the water pick-up kinetics in polymeric composites in a complex method [81]. However, higher segmental movement of the polymeric chains speeds up the water intake rate at higher conditioning temperatures. In contrary to low-temperature water aging, high temperature conditioning (HTC) shows the harsh and hostile influence of MWCNTs in the glass fiber/epoxy (GE) (Figure 5.12b) composite. It can be noticed that a higher water uptake rate, as well as greater saturation content, can be seen for a MWCNT-GE composite as compared to a control GE composite. The water diffusivity for GE and MWCNT-GE composites at this temperature was found to be 2.3×10^{-6} and 2.7×10^{-6} mm^2/s (Figure 5.12d), respectively. Further, it can also be observed that a deviation from linearity between M_t and $\sqrt{t}$ is found at a much lower conditioning time in HTC than LTC. Conditioning to higher temperature, water generates a thermal stress at the MWCNT/matrix interface due to a differential coefficient of thermal expansion of the MWCNTs ($0.73 - 1.49 \times 10^{-5}$ K^{-138}) and polymer epoxy (6.2×10^{-5} K^{-139}) [79]. As the epoxy matrix is likely to swell radially at a higher rate than MWCNTs, tensile stress is possible to be generated near the interphase polymer area, provided the MWCNTs and epoxy are well bonded at the interface. Therefore, there is a better chance of MWCNT/polymer interfacial slippage at higher temperatures [82,83]. Furthermore, the superior attraction of epoxy to absorb more water content above MWCNTs leads to the initiation of hygroscopic stress at the MWCNT/epoxy interface. This stressed interface fails to check the mobility of water efficiently. Extended exposure of specimens to high-temperature water may ultimately lead to various intermittently micron-scaled debonded regions, through which water induced by capillary action becomes accelerated [84].

Figure 5.13 depicts the flexural behavior of control GFRP and MWCNT-GFRP composites at different durations of water aging at the time of LTC and HTC. During conditioning time, the MWCNT/GFRP composite exhibited 24% higher strength and modulus above the control GFRP composite. The possible reason for enhanced mechanical properties may be governed by good dispersion of MWCNTs in the polymer region. This possibly produces higher interfacial area and that facilitates the smooth transfer of stress from the matrix to the MWCNT nanoparticles. In the case of LTC for a different time duration reveals a minor effect on the flexural strength value, the modulus value has been dropped by 11% and 14% at the saturation point for the neat GFRP and MWCNT/GFRP composites as compared to the before conditioning-tested specimen. This denotes a higher amount of plasticization in

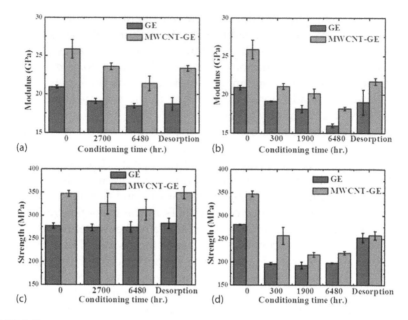

FIGURE 5.13
Flexural properties of GE and MWCNT-GE composites after (a, b) LTC and (c, d) HTC for various time intervals and after complete desorption. (From Prusty, R.K. et al., *J. Appl. Polym. Sci.*, 135, 45987, 2018.)

the MWCNT/GFRP composites. Interestingly, at this period, MWCNT/GFRP composites showed a near about 20% higher modulus value as compared to the neat GFRP composite. An insignificant recovery in the modulus value of both the composites was seen after complete desorption. Whereas, in the case of HTC, both strength and modulus value were constantly decreased with increasing condition time. During the saturation time, the strength value of the GFRP and MWCNT/GFRP composites was decreased by 30% and 37%, respectively, with respect to the before conditioning stage. The modulus value of the GFRP and MWCNT/GFRP composites showed a decrement of 24% and 30%, respectively, as compared to the before conditioning stage. This displays a higher and detrimental amount of hydrolysis and plasticization in the composites at the time that HTC attributed to accelerated water ingression. This higher degree of damage and degradation in the mechanical properties in MWCNT/GFRP composites may be clarified in regards to the higher percentage of water and irreversible interfacial damage due to long-term high-temperature water conditioning of this composite. Meanwhile, at this phase, the MWCNT/GFRP composite showed an improvement in the strength and modulus value of 11% and 14%, respectively, as compared to the neat GFRP composite.

After the complete saturation and desorption of both the composites systems, the immersed water causes substantial recovery in mechanical properties due to the reversible nature of hydrolysis of the polymer phase. In the case of the HTC specimen, after complete desorption shows a greater decrease in the strength and modulus value of 10% and 9%, respectively, for the GFRP composite than that of the before-conditioning stage. Similarly, for the MWCNT/GFRP composite the strength and modulus value decreased by 25% and 15%, respectively. The above exploration in a hydrothermal environment leads to a greater amount of irreversible damage and degradation in mechanical properties of the MWCNT/GFRP composite than the GFRP composite. Due to the propagation of thermal stresses at a higher temperature and the generation of hygroscopic stress due to differential water absorption, the development of interfacial debonding at the MWCNT/polymer interface is quickened. This type of irreparable interfacial damage and degradation may be attributed to the lower degree of recovery in terms of mechanical properties of the MWCNT/GFRP composite as compared to the neat GFRP composite in their desorbed phase.

Gkikas et al. [85] have investigated the effect of hygrothermal conditioning on MWCNTs-modified epoxy-based polymeric systems. 0.5% MWCNT-embedded composites have revealed neither any impact on the moisture uptake kinetic of the material nor on the saturation moisture content. Water diffusivity was found to be 0.0438 mm^2/h for both the cases, i.e., the neat epoxy and MWCNT/epoxy composites. But, in the CNT-enhanced CFRP composite, a substantial change in equilibrium water content and moisture uptake behavior was observed. The neat CFRP composite exhibited a maximum water uptake rate as compared to other MWCNT-modified composites. The equilibrium water concentration in the neat CFRP and MWCNT-enhanced CFRP composites was 1% and 0.6%, respectively. The equilibrium water content and diffusion coefficient for the doped CFRP composite were 40% and 10% lower than the neat CFRP composite. Imperfect polymerization in the CFRP composite could be one of the possible results in the loose packing of the polymeric chains and smaller molecules, which creates higher free volume in the polymeric composite. This higher free volume generates active direction for moisture ingression. In doped CFRP, the MWCNTs may be adjusted in the existing free volume and therefore makes the composite extra-rigid and dense, which repels moisture ingression.

Garg et al. [86] have performed experiments on the effects of pristine and amino-functionalized CNTs on the water absorption and residual mechanical properties of the glass FRP composites. They found a substantial decrease in the diffusion coefficient and the saturation moisture content in the existence of CNTs. Furthermore, the amino-functionalization of CNTs provides improved resistance to water uptake than pristine CNTs. The investigation showed that the material degradation of the composites is less due to the presence of CNTs in seawater as compared to distilled water.

5.6 Synergistic Effect of Nanofillers

In this section, the combined effect of two or more nanofillers in the polymeric composites in the hydrothermal environment has been discussed. CNTs and GNPs have several similar properties; however, they vary in other features due to the structural differences. Experiments have recently been made to incorporate these two carbon-based materials in order to exploit the advantages of both [87]. CNTs have excellent electrical conductivity, and they can be added to a poor electroconductive polymer to increase its electrical properties [88]. Similarly, GNPs have outstanding thermal conductivity of 3000 W/mK, and they can well increase the thermal conductivity of a polymer [31].

Moosa et al. [89] investigated the combined effects of GNPs and nonfunctionalized MWCNTs hybrid nanofiller and their dispersions on mechanical, wettability, and thermal stability properties of epoxy nanocomposites. The tensile strength of epoxy/GNPs-MWCNTs hybrid nano-composites containing a fixed weight fraction of GNPs/MWCNTs hybrids (0.5 wt%) with various mixing ratios (0:10, 1:9, 2.5:7.5, 5:5, 7.5:2.5, 9:1, and 10:0) of GNPs/MWCNTs are revealed in Figure 5.14. The increases in tensile strength are 27%, 46%, 50%, 53%, 52%, 60%, and 59% at GNPs/MWCNTs mixing ratios of 0:10, 1:9, 2.5:7.5, 5:5, 7.5:2.5, 9:1, and 10:0, respectively. The tensile strength with a mixing ratio of (9:1) GNPs/MWCNTs was exhibited as higher among all the other composites than that for the neat epoxy or single type of carbon nanomaterial-reinforced epoxy (0:10 and 10:0). This may be attributed to the flexible one-dimensional structure MWCNTs with the two-dimensional structure GNPs that can form three-dimensional hybrid architectures. These three-dimensional architectures will prevent face to face accumulation of GNPs. This results in a higher contact surface area between the GNPs/MWCNTs and epoxy matrix and

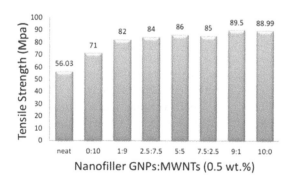

FIGURE 5.14
Tensile strength of epoxy/GNPs-MCWNTs nano-composites. (From Moosa, A.A. et al., *Am. J. Mater. Sci.*, 7, 1–11, 2017.)

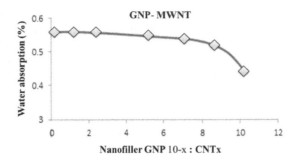

FIGURE 5.15
Water absorption vs. GNPs-MWNTs nanofillers with fixed weight fraction (0.5 wt%) at different mixing ratios. (From Moosa, A.A. et al., *Am. J. Mater. Sci.*, 7, 1–11, 2017.)

also MWCNTs acted as extended synchronizing supports for the three-dimensional hybrid architectures, which can become entangled with the polymeric chain resulting in superior interface between GNPs/MWCNTs hybrid and the epoxy matrix [90,91].

Figure 5.15 depicts the water absorption percentage of the epoxy/GNPs/MWCNTs hybrid nano-composite specimens with a fixed nanofiller weight fraction (0.5 wt%) and a different mixing ratio. The water absorbance behavior of the epoxy/GNP-MWCNT nano-composite is smaller than that of the epoxy/GNP nano-composite and higher than that of the epoxy/MWCNTs composites. This may be attributed to the GNPs natural propensity to absorb water more than the MWCNTs and also due to a higher exposed surface area of the nanoplatelets [92]. It is evident that the epoxy/MWCNTs nano-composite is more hydrophobic than the epoxy/GNP/MWCNT and epoxy/GNPs nano-composite, which is described by contact angle measurement. In the meantime, GNPs generated a tortuous pathway in the epoxy to defuse the water molecule inside the nano-composite because it has a high aspect ratio [93].

References

1. Kulpinski, P. Cellulose fibers modified by silicon dioxide nanoparticles. *Journal of Applied Polymer Science* 98 (2005): 1793–1798. doi:10.1002/app.22279.
2. Kedem, S., J. Schmidt, Y. Paz, and Y. Cohen. Composite polymer nanofibers with carbon nanotubes and titanium dioxide particles. *Langmuir* 21 (2005): 5600–5604. doi:10.1021/la0502443.
3. Zelikman, E., R. Y. Suckeveriene, G. Mechrez, and M. Narkis. Fabrication of composite polyaniline/CNT nanofibers using an ultrasonically assisted dynamic inverse emulsion polymerization technique. *Polymers of Advanced Technologies* 21 (2010): 150–152. doi:10.1002/pat.1464.

4. Gao, J., M. E. Itkis, A. Yu, E. Bekyarova, B. Zhao, and R. C. Haddon. Continuous spinning of a single-walled carbon nanotube–nylon composite fiber. *Journal of the American Chemical Society* 127 (2005): 3847–3854. doi:10.1021/ja0446193.

5. Perreault, F., A. F. de Faria, and M. Elimelech. Environmental applications of graphene-based nanomaterials. *Chemical Society Reviews* 44 (2015): 5861–5896. doi:10.1039/C5CS00021A.

6. Zhou, J., and J. P. Lucas. The effects of a water environment on anomalous absorption behavior in graphite/epoxy composites. *Composites Science and Technology* 53 (1995): 57–64. doi:10.1016/0266-3538(94)00078-6.

7. Zhou, J., and J. P. Lucas. Hygrothermal effects of epoxy resin. Part II: Variations of glass transition temperature. *Polymer* 40 (1999): 5513–5522. doi:10.1016/S0032-3861(98)00791-5.

8. Zhou, J., and J. P. Lucas. Hygrothermal effects of epoxy resin. Part I: The nature of water in epoxy. *Polymer* 40 (1999): 5505–5512. doi:10.1016/S0032-3861(98)00790-3.

9. Lin, Y. C., and X. Chen. Investigation of moisture diffusion in epoxy system: Experiments and molecular dynamics simulations. *Chemical Physics Letters* 412 (2005): 322–326. doi:10.1016/j.cplett.2005.07.022.

10. Popineau, S., C. Rondeau-Mouro, C. Sulpice-Gaillet, and M. E. R. Shanahan. Free/bound water absorption in an epoxy adhesive. *Polymer* 46 (2005): 10733–10740. doi:10.1016/j.polymer.2005.09.008.

11. Wu, C., and W. Xu. Atomistic simulation study of absorbed water influence on structure and properties of crosslinked epoxy resin. *Polymer* 48 (2007): 5440–5448. doi:10.1016/j.polymer.2007.06.038.

12. Prolongo, S. G., M. R. Gude, and A. Ureña. Water uptake of epoxy composites reinforced with carbon nanofillers. *Composites Part A: Applied Science and Manufacturing* 43 (2012): 2169–2175. doi:10.1016/j.compositesa.2012.07.014.

13. Shen, C.-H., and G. S. Springer. Moisture absorption and desorption of composite materials. *Journal of Composite Materials* 10 (1976): 2–20. doi:10.1177/002199837601000101.

14. Cai, L.-W., and Y. Weitsman. Non-Fickian moisture diffusion in polymeric composites. *Journal of Composite Materials* 28 (1994): 130–154. doi:10.1177/002199839402800203.

15. Roy, S., W. X. Xu, S. J. Park, and K. M. Liechti. Anomalous moisture diffusion in viscoelastic polymers: Modeling and testing. *Journal of Applied Mechanics* 67 (2000): 391–396. doi:10.1115/1.1304912.

16. Marín, L., E. V. González, P. Maimí, D. Trias, and P. P. Camanho. Hygrothermal effects on the translaminar fracture toughness of cross-ply carbon/epoxy laminates: Failure mechanisms. *Composites Science and Technology* 122 (2016): 130–139. doi:10.1016/j.compscitech.2015.10.020.

17. Prusty, R. K., D. K. Rathore, and B. C. Ray. CNT/polymer interface in polymeric composites and its sensitivity study at different environments. *Advances in Colloid and Interface Science* 240 (2017): 77–106. doi:10.1016/j.cis.2016.12.008.

18. Van Noorden, R. Production: Beyond sticky tape. *Nature* 483 (2012): S32–S33. doi:10.1038/483S32a.

19. Boehm, H.-P. Graphene—How a laboratory curiosity suddenly became extremely interesting. *Angewandte Chemie International Edition* 49 (2010): 9332–9335. doi:10.1002/anie.201004096.

20. Novoselov, K. S., A. K. Geim, S. V. Morozov, D. Jiang, Y. Zhang, S. V. Dubonos, I. V. Grigorieva, and A. A. Firsov. Electric field effect in atomically thin carbon films. *Science* 306 (2004): 666–669. doi:10.1126/science.1102896.

21. Nair, R. R., H. A. Wu, P. N. Jayaram, I. V. Grigorieva, and A. K. Geim. Unimpeded permeation of water through helium-leak–tight graphene-based membranes. *Science* 335 (2012): 442–444. doi:10.1126/science.1211694.

22. Geim, A. K., and K. S. Novoselov. The rise of graphene. *Nature Materials* 6 (2007): 183–191. doi:10.1038/nmat1849.

23. Pan, Y., N. G. Sahoo, and L. Li. The application of graphene oxide in drug delivery, *Expert Opinion on Drug Delivery* 9 (2012): 1365–1376. doi:10.1517/17425247.2012.729575.

24. Service, R. F. Carbon sheets an atom thick give rise to graphene dreams. *Science* 324 (2009): 875–877. doi:10.1126/science.324_875.

25. Lachman, N., C. Bartholome, P. Miaudet, M. Maugey, P. Poulin, and H. D. Wagner. Raman response of carbon nanotube/PVA fibers under strain. *Journal of Physical Chemistry C* 113 (2009): 4751–4754. doi:10.1021/jp900355k.

26. Dalton, A. B., S. Collins, E. Muñoz, J. M. Razal, V. H. Ebron, J. P. Ferraris, J. N. Coleman, B. G. Kim, and R. H. Baughman. Super-tough carbon-nanotube fibres. *Nature* 423 (2003): 703–703. doi:10.1038/423703a.

27. Shin, M. K., B. Lee, S. H. Kim, J. A. Lee, G. M. Spinks, S. Gambhir, G. G. Wallace et al. Synergistic toughening of composite fibres by self-alignment of reduced graphene oxide and carbon nanotubes. *Nature Communications* 3 (2012): 650. doi:10.1038/ncomms1661.

28. Ren, G., Z. Zhang, X. Zhu, B. Ge, F. Guo, X. Men, and W. Liu. Influence of functional graphene as filler on the tribological behaviors of Nomex fabric/phenolic composite. *Composites Part A: Applied Science and Manufacturing* 49 (2013): 157–164. doi:10.1016/j.compositesa.2013.03.001.

29. Xu, Z., and C. Gao. In situ polymerization approach to graphene-reinforced nylon-6 composites. *Macromolecules* 43 (2010): 6716–6723. doi:10.1021/ma1009337.

30. Ajayan, P. M., and J. M. Tour. Materials science: Nanotube composites. *Nature* 447 (2007): 1066–1068. doi:10.1038/4471066a.

31. Stankovich, S., D. A. Dikin, G. H. B. Dommett, K. M. Kohlhaas, E. J. Zimney, E. A. Stach, R. D. Piner, S. T. Nguyen, and R. S. Ruoff. Graphene-based composite materials. *Nature* 442 (2006): 282–286. doi:10.1038/nature04969.

32. Lee, C., X. Wei, J. W. Kysar, and J. Hone. Measurement of the elastic properties and intrinsic strength of monolayer graphene. *Science* 321 (2008): 385–388. doi:10.1126/science.1157996.

33. Gómez-Navarro, C., M. Burghard, and K. Kern. Elastic properties of chemically derived single graphene sheets. *Nano Letters* 8 (2008): 2045–2049. doi:10.1021/nl801384y.

34. Wu, Y., M. Chen, M. Chen, Z. Ran, C. Zhu, and H. Liao. The reinforcing effect of polydopamine functionalized graphene nanoplatelets on the mechanical properties of epoxy resins at cryogenic temperature. *Polymer Testing* 58 (2017): 262–269. doi:10.1016/j.polymertesting.2016.12.021.

35. Sun, X., X. Liu, X. Shen, Y. Wu, Z. Wang, and J.-K. Kim. Graphene foam/carbon nanotube/poly(dimethyl siloxane) composites for exceptional microwave shielding. *Composites Part Applied Science and Manufacturing* 85 (2016): 199–206. doi:10.1016/j.compositesa.2016.03.009.

36. Huang, C. J., S. Y. Fu, Y. H. Zhang, B. Lauke, L. F. Li, and L. Ye. Cryogenic properties of SiO_2/epoxy nanocomposites. *Cryogenics* 45 (2005): 450–454. doi:10.1016/j.cryogenics.2005.03.003.
37. Ashori, A., H. Rahmani, and R. Bahrami. Preparation and characterization of functionalized graphene oxide/carbon fiber/epoxy nanocomposites. *Polymer Testing* 48 (2015): 82–88. doi:10.1016/j.polymertesting.2015.09.010.
38. Low, F. W., C. W. Lai, and S. B. Abd Hamid. Facile synthesis of high quality graphene oxide from graphite flakes using improved Hummer's technique. *Journal of Nanoscience and Nanotechnology* 15 (2015): 6769–6773. doi:10.1166/jnn.2015.10903.
39. Li, Z., R. Wang, R. J. Young, L. Deng, F. Yang, L. Hao, W. Jiao, and W. Liu. Control of the functionality of graphene oxide for its application in epoxy nanocomposites. *Polymer* 54 (2013): 6437–6446. doi:10.1016/j.polymer.2013.09.054.
40. George, J. J., A. Bandyopadhyay, and A. K. Bhowmick. New generation layered nanocomposites derived from ethylene-co-vinyl acetate and naturally occurring graphite. *Journal of Applied Polymer Science* 108 (2008): 1603–1616. doi:10.1002/app.25067.
41. Wang, C., J. Li, S. Sun, X. Li, F. Zhao, B. Jiang, and Y. Huang. Electrophoretic deposition of graphene oxide on continuous carbon fibers for reinforcement of both tensile and interfacial strength. *Composites Science and Technology* 135 (2016): 46–53. doi:10.1016/j.compscitech.2016.07.009.
42. Hung, P., K. Lau, B. Fox, N. Hameed, J. H. Lee, and D. Hui. Surface modification of carbon fibre using graphene–related materials for multifunctional composites. *Composites Part B: Engineering* 133 (2018): 240–257. doi:10.1016/j.compositesb.2017.09.010.
43. Coleman, E. A. 23-Plastics additives. In M. Kutz (Ed.), *Applied Plastics Engineering Handbook*. William Andrew Publishing, Oxford, 2011: pp. 419–428. doi:10.1016/B978-1-4377-3514-7.10023-6.
44. Campbell, F. C. Chapter 2-Fibers and reinforcements: The string that provides the strength. In F. C. Campbell (Ed.), *Manufacturing Processes for Advanced Composites*, Elsevier Science, Amsterdam, the Netherlands, 2004: pp. 39–62. doi:10.1016/B978-185617415-2/50003-4.
45. Campbell, F. C. *Structural Composite Materials*, ASM International, 2010.
46. Zhou, S., S. Chiang, J. Xu, H. Du, B. Li, C. Xu, and F. Kang. Modeling the in-plane thermal conductivity of a graphite/polymer composite sheet with a very high content of natural flake graphite. *Carbon* 50 (2012): 5052–5061. doi:10.1016/j.carbon.2012.06.045.
47. Su, J., and J. Zhang. Improvement of mechanical and dielectrical properties of ethylene propylene diene monomer (EPDM)/barium titanate ($BaTiO_3$) by layered mica and graphite flakes. *Composites Part B: Engineering* 112 (2017): 148–157. doi:10.1016/j.compositesb.2017.01.002.
48. Xue, C., H. Bai, P. F. Tao, N. Jiang, and S. L. Wang. Analysis on thermal conductivity of graphite/Al composite by experimental and modeling study. *Journal of Materials Engineering and Performance* 26 (2017): 327–334. doi:10.1007/s11665-016-2447-z.
49. Ahmad, H., M. Fan, and D. Hui. Graphene oxide incorporated functional materials: A review. *Composites Part B: Engineering* 145 (2018): 270–280. doi:10.1016/j.compositesb.2018.02.006.

50. Kim, K.-S., K.-Y. Rhee, K.-H. Lee, J.-H. Byun, and S.-J. Park. Rheological behaviors and mechanical properties of graphite nanoplate/carbon nanotube-filled epoxy nanocomposites. *Journal of Industrial and Engineering Chemistry* 16 (2010): 572–576. doi:10.1016/j.jiec.2010.03.017.

51. Zhou, S., Y. Chen, H. Zou, and M. Liang. Thermally conductive composites obtained by flake graphite filling immiscible polyamide 6/polycarbonate blends. *Thermochimica Acta* 566 (2013): 84–91. doi:10.1016/j.tca.2013.05.027.

52. Su, Z., H. Wang, K. Tian, W. Huang, Y. Guo, J. He, and X. Tian. Multifunctional anisotropic flexible cycloaliphatic epoxy resin nanocomposites reinforced by aligned graphite flake with non-covalent biomimetic functionalization. *Composites Part A: Applied Science and Manufacturing* 109 (2018): 472–480. doi:10.1016/j.compositesa.2018.02.033.

53. Zhou, C., W. Huang, Z. Chen, G. Ji, M. L. Wang, D. Chen, and H. W. Wang. In-plane thermal enhancement behaviors of Al matrix composites with oriented graphite flake alignment. *Composites Part B: Engineering* 70 (2015): 256–262. doi:10.1016/j.compositesb.2014.11.018.

54. Sun, K., Y. Qiu, and L. Zhang. Preserving flake size in an African flake graphite ore beneficiation using a modified grinding and pre-screening process. *Minerals* 7 (2017): 115. doi:10.3390/min7070115.

55. Kim, S. H., Y.-J. Heo, M. Park, B.-G. Min, K. Y. Rhee, and S.-J. Park. Effect of hydrophilic graphite flake on thermal conductivity and fracture toughness of basalt fibers/epoxy composites. *Composites Part B: Engineering* 153 (2018): 9–16. doi:10.1016/j.compositesb.2018.07.022.

56. Kostagiannakopoulou, C., X. Tsilimigkra, G. Sotiriadis, and V. Kostopoulos. Synergy effect of carbon nano-fillers on the fracture toughness of structural composites. *Composites Part B: Engineering* 129 (2017): 18–25. doi:10.1016/j.compositesb.2017.07.012.

57. Zheng, S., J. Wang, Q. Guo, J. Wei, and J. Li. Miscibility, morphology and fracture toughness of epoxy resin/poly(styrene-co-acrylonitrile) blends. *Polymer* 37 (1996): 4667–4673. doi:10.1016/S0032-3861(96)00324-2.

58. Cha, J., G. H. Jun, J. K. Park, J. C. Kim, H. J. Ryu, and S. H. Hong. Improvement of modulus, strength and fracture toughness of CNT/Epoxy nanocomposites through the functionalization of carbon nanotubes. *Composites Part B: Engineering* 129 (2017): 169–179. doi:10.1016/j.compositesb.2017.07.070.

59. Demirci, M. T., N. Tarakçıoğlu, A. Avcı, A. Akdemir, and İ. Demirci. Fracture toughness (Mode I) characterization of SiO$_2$ nanoparticle filled basalt/epoxy filament wound composite ring with split-disk test method. *Composites Part B: Engineering* 119 (2017): 114–124. doi:10.1016/j.compositesb.2017.03.045.

60. Yeo, J.-S., O. Y. Kim, and S.-H. Hwang. The effect of chemical surface treatment on the fracture toughness of microfibrillated cellulose reinforced epoxy composites. *Journal of Industrial and Engineering Chemistry* 45 (2017): 301–306. doi:10.1016/j.jiec.2016.09.039.

61. Janas, D. Towards monochiral carbon nanotubes: A review of progress in the sorting of single-walled carbon nanotubes. *Materials Chemistry Frontiers* 2 (2017): 36–63. doi:10.1039/C7QM00427C.

62. Hodge, S. A., M. K. Bayazit, K. S. Coleman, and M. S. P. Shaffer. Unweaving the rainbow: A review of the relationship between single-walled carbon nanotube molecular structures and their chemical reactivity. *Chemical Society Reviews* 41 (2012): 4409–4429. doi:10.1039/c2cs15334c.

63. Hong, S., and S. Myung. Nanotube electronics: A flexible approach to mobility. *Nature Nanotechnology* 2 (2007): 207–208. doi:10.1038/nnano.2007.89.
64. Brady, G. J., A. J. Way, N. S. Safron, H. T. Evensen, P. Gopalan, and M. S. Arnold. Quasi-ballistic carbon nanotube array transistors with current density exceeding Si and GaAs. *Science Advances* 2 (2016): e1601240. doi:10.1126/sciadv.1601240.
65. Janas, D., A. P. Herman, S. Boncel, and K. K. K. Koziol. Iodine monochloride as a powerful enhancer of electrical conductivity of carbon nanotube wires. *Carbon* 73 (2014) 225–233. doi:10.1016/j.carbon.2014.02.058.
66. Yu, L., C. Shearer, and J. Shapter. Recent development of carbon nanotube transparent conductive films. *Chemical Reviews* 116 (2016): 13413–13453. doi:10.1021/acs.chemrev.6b00179.
67. Koziol, K. K., D. Janas, E. Brown, and L. Hao. Thermal properties of continuously spun carbon nanotube fibres. *Physica E: Low-Dimensional Systems and Nanostructures* 88 (2017): 104–108. doi:10.1016/j.physe.2016.12.011.
68. Pop, E., D. Mann, Q. Wang, K. Goodson, and H. Dai. Thermal conductance of an individual single-wall carbon nanotube above room temperature. *Nano Letters* 6 (2006): 96–100. doi:10.1021/nl052145f.
69. Han, Z., and A. Fina. Thermal conductivity of carbon nanotubes and their polymer nanocomposites: A review. *Progress in Polymer Science* 36 (2011): 914–944. doi:10.1016/j.progpolymsci.2010.11.004.
70. Dresselhaus, M. S., G. Dresselhaus, and P. Avouris, Eds. *Carbon Nanotubes: Synthesis, Structure, Properties, and Applications*. Springer-Verlag, Berlin, Germany, 2001. https://www.springer.com/gp/book/9783540410867 (Accessed August 22, 2019).
71. Janas, D., N. Czechowski, B. Krajnik, S. Mackowski, and K. K. Koziol. Electroluminescence from carbon nanotube films resistively heated in air. *Applied Physics Letters* 102 (2013): 181104. doi:10.1063/1.4804296.
72. Bachilo, S. M., M. S. Strano, C. Kittrell, R. H. Hauge, R. E. Smalley, and R. B. Weisman. Structure-assigned optical spectra of single-walled carbon nanotubes. *Science* 298 (2002): 2361–2366. doi:10.1126/science.1078727.
73. Star, A., Y. Lu, K. Bradley, and G. Grüner. Nanotube optoelectronic memory devices. *Nano Letters* 4 (2004): 1587–1591. doi:10.1021/nl049337f.
74. Yu, M.-F., O. Lourie, M. J. Dyer, K. Moloni, T. F. Kelly, and R. S. Ruoff. Strength and breaking mechanism of multiwalled carbon nanotubes under tensile load. *Science* 287 (2000): 637–640. doi:10.1126/science.287.5453.637.
75. Li, Q., M. Zaiser, J. R. Blackford, C. Jeffree, Y. He, and V. Koutsos. Mechanical properties and microstructure of single-wall carbon nanotube/elastomeric epoxy composites with block copolymers. *Materials Letters* 125 (2014): 116–119. doi:10.1016/j.matlet.2014.03.096.
76. Yu, M. F., B. S. Files, S. Arepalli, and R. S. Ruoff. Tensile loading of ropes of single wall carbon nanotubes and their mechanical properties. *Physical Review Letters* 84 (2000): 5552–5555. doi:10.1103/PhysRevLett.84.5552.
77. Zulfli, N. M., A. A. Bakar, and W. Chow. Mechanical and water absorption behaviors of carbon nanotube reinforced epoxy/glass fiber laminates. *Journal of Reinforced Plastics and Composites* 32 (2013): 1715–1721. doi:10.1177/0731684413501926.

78. Starkova, O., S. T. Buschhorn, E. Mannov, K. Schulte, and A. Aniskevich. Water transport in epoxy/MWCNT composites. *European Polymer Journal* 49 (2013): 2138–2148. doi:10.1016/j.eurpolymj.2013.05.010.

79. Prusty, R. K., D. K. Rathore, and B. C. Ray. Water-induced degradations in MWCNT embedded glass fiber/epoxy composites: An emphasis on aging temperature. *Journal of Applied Polymer Science* 135 (2018): 45987. doi:10.1002/app.45987.

80. Cangialosi, D., V. M. Boucher, A. Alegría, and J. Colmenero. Enhanced physical aging of polymer nanocomposites: The key role of the area to volume ratio. *Polymer* 53 (2012): 1362–1372. doi:10.1016/j.polymer.2012.01.033.

81. Ray, B. C., and D. Rathore. Environmental damage and degradation of FRP composites: A review report. *Polymer Composites* 36 (2015): 410–423. doi:10.1002/pc.22967.

82. Zhang, Y. C., and X. Wang. Hygrothermal effects on interfacial stress transfer characteristics of carbon nanotubes-reinforced composites system. *Journal of Reinforced Plastics and Composites* (2016). doi:10.1177/0731684406055456.

83. Suhr, J., W. Zhang, P. M. Ajayan, and N. A. Koratkar. Temperature-activated interfacial friction damping in carbon nanotube polymer composites. *Nano Letters* 6 (2006): 219–223. doi:10.1021/nl0521524.

84. Ray, B. C. Temperature effect during humid ageing on interfaces of glass and carbon fibers reinforced epoxy composites. *Journal of Colloid and Interface Science* 298 (2006): 111–117. doi:10.1016/j.jcis.2005.12.023.

85. Gkikas, G., D.-D. Douka, N.-M. Barkoula, and A. S. Paipetis. Nano-enhanced composite materials under thermal shock and environmental degradation: A durability study. *Composites Part B: Engineering* 70 (2015): 206–214. doi:10.1016/j.compositesb.2014.11.008.

86. Garg, M., S. Sharma, and R. Mehta. Carbon nanotube-reinforced glass fiber epoxy composite laminates exposed to hygrothermal conditioning. *Journal of Materials Science* 51 (2016): 8562–8578. doi:10.1007/s10853-016-0117-z.

87. Pu, J., S. Wan, Z. Lu, G. Zhang, L. Wang, X. Zhang, and Q. Xue. Controlled water adhesion and electrowetting of conducting hydrophobic graphene/carbon nanotubes composite films on engineering materials. *Journal of Materials Chemistry A* 1 (2012): 1254–1260. doi:10.1039/C2TA00344A.

88. Ebbesen, T. W., H. J. Lezec, H. Hiura, J. W. Bennett, H. F. Ghaemi, and T. Thio. Electrical conductivity of individual carbon nanotubes. *Nature* 382 (1996): 54–56. doi:10.1038/382054a0.

89. Moosa, A. A., S. A. A. Ramazani, F. A. K. Kubba, and M. Raad. Synergetic effects of graphene and nonfunctionalized carbon nanotubes hybrid reinforced epoxy matrix on mechanical, thermal and wettability properties of nanocomposites. *American Journal of Materials Science* 7 (2017): 1–11.

90. Yang, S.-Y., W.-N. Lin, Y.-L. Huang, H.-W. Tien, J.-Y. Wang, C.-C. M. Ma, S.-M. Li, and Y.-S. Wang. Synergetic effects of graphene platelets and carbon nanotubes on the mechanical and thermal properties of epoxy composites. *Carbon* 49 (2011): 793–803. doi:10.1016/j.carbon.2010.10.014.

91. Wang, P.-N., T.-H. Hsieh, C.-L. Chiang, and M.-Y. Shen. Synergetic effects of mechanical properties on graphene nanoplatelet and multiwalled carbon nanotube hybrids reinforced epoxy/carbon fiber composites. *Journal of Nanomaterials* (2015). doi:10.1155/2015/838032.

92. Zaman, I., S. Araby, A. Khalid, and B. Manshoor. Moisture absorption and diffusivity of epoxy filled layered-structure nanocomposite. *Advances in Environmental Biology* (2014). https://link.galegroup.com/apps/doc/A392070199/AONE?sid=lms (Accessed August 24, 2019).
93. Zhao, H., and R. K. Y. Li. Effect of water absorption on the mechanical and dielectric properties of nano-alumina filled epoxy nanocomposites. *Composites Part A: Applied Science and Manufacturing* 39 (2008): 602–611. doi:10.1016/j.compositesa.2007.07.006.

6

Hydrothermal Effect on Mechanical Properties of Inorganic Nanofillers Embedded FRP Composites

6.1 Introduction

Fiber reinforced polymeric (FRP) composites nowadays have become one of the suitable choices of materials in terms of fabricating and designing for a wide range of engineering applications owing to their high specific strength and stiffness, excellent corrosion resistance, and compatibility with other materials as compared to their counterparts. In service conditions, these materials were affected by numerous environmental parameters such as moisture, seawater, hydrothermal, alkaline, UV light, and other corrosive environments. Therefore, it is important to limit the long-term exposure of these low density materials in regard to their durability and reliability. The micro/nano inorganic particles are incorporated into the epoxy matrix to increase the overall mechanical properties of the composites, which can broaden the areas of application of the materials. Inorganic nanofillers are more suitable as potential fillers because of their low cost and ease of fabrication technique [1]. Various researchers and scientists have experimentally investigated that the incorporation of inorganic micro/nanofillers into the polymeric matrix certainly enhances the mechanical, wear, thermomechanical, Young's modulus, impact, fracture toughness, thermal conductivity, and glass transition temperature.

However, these FRPs-based materials are encountering threats and challenges at diverse environmental conditions like temperature variations (low, high, thermal shock, thermal spike) and alkaline, water, corrosive, and UV light exposure. Usually, in a high relative humidity/moisture or hydrothermal environment, polymeric composites are subjected to absorb moisture followed by damage and degradation of their thermal, mechanical, electrical, and physical behavior. In the case of an epoxy polymer, the water molecules/absorbed moisture are generally attached with the hydroxyl group of epoxy and the free water is gathered in the voids/free volume existing inside the polymer or at the region of the fiber/matrix interface [2,3]. Water absorbed by the polymers changes their properties both chemically and physically. The chemical modifications of the polymer epoxy are chain hydrolysis and scission [4–7]. The physical variations of the polymer are basically due to plasticization and swelling. The glass transition temperature (T_g) behavior of FRP composites is affected by a physical change in the structure of the polymer. The overall mechanical behavior is governed by the chemical and physical behavior and the structure of the epoxy polymer. Generally, moisture absorption changes the chemical, thermo-physical, and mechanical characteristics of FRP composites [8]. An additional phase of the damage and degradation of mechanical behavior is the generation of the microcracking in the matrix under different environmental conditions [9–11]. Therefore, maintaining the mechanical behavior in a hydrothermal environment is crucial and challenging to the researchers and design engineers.

The usage of nanoparticles has been widely progressed with polymer matrix composites (PMCs) as reinforcements as compared to their counterparts such as microparticles. It is of utmost importance to improve the mechanical properties of these polymeric-based composites using the inclusion of smaller volume fractions of inorganic nanofillers such as SiO_2, Al_2O_3, TiO_2, SiC, $CaCO_3$, BN, as well as the combination of the different inorganic nanofillers. The incorporation of about 1%–5% inorganic nanofillers (TiO_2, Al_2O_3, SiO_2, SiC, $CaCO_3$, BN, etc.) in the polymeric phase definitely increases the mechanical properties of the PMC [12–18]. Because of the tiny size of the particles and the absence of any larger aggregates, the nanofillers can easily infiltrate to all polymeric and fiber structures without compromising the impregnation by excessive viscosity, thereby enabling all the state-of-the-art process technologies like resin infusion, resin transfer molding (RTM), or resin injection to be used [19].

One of the possible approaches to increase the interface strength is by the incorporation of nanofillers into FRP composites [14,20–23]. Okhawilai et al. [24] have studied the effect of nano-SiO_2 on the epoxy polymer, and they reported that there is an enhancement of modulus by 2.5 times with the incorporation of nano-SiO_2 particles in epoxy-modified polybenzoxazine. Inorganic particles/fillers are mixed with polymeric materials either in dispersed form, mechanically contacted, chemically bonded, or a combination of two or all of these [25,26]. Microcracks may bend, pin, or blunt at the nanoparticle surface during their transmission, resulting in the enhancement of fracture toughness [27]. Therefore, the strength of the interface or interphase governs the durability and reliability of the composites at different environmental conditions. It has been experimentally seen that organic nanofillers like SiO_2, Al_2O_3, and TiO_2 decrease water affinity and permeability, causing the enhancement of corrosive resistivity and hydrolytic degradation of FRP composites [17,28,29]. Though research on water absorption behavior for FRP composites is not new, the investigation of residual mechanical properties of inorganic nanofillers-filled FRP composites that are exposed to the acceleration of hydrothermal aging is rather exciting and essential. Therefore, an effort has been made to study the effect of inorganic nanofiller concentration on the water absorption and residual mechanical and thermal properties of FRP composites. The moisture diffusion kinetics have been presented with a comparison between the raw and the nanofiller-enhanced FRP composites. Residual thermal and mechanical behavior also have been articulated for hydrothermally conditioned FRP composites. Further, micro and nano dimension failure modes and strengthening mechanisms of the polymeric composites have been presented in the current discussion.

6.2 SiO$_2$

Gonon et al. investigated the effects of hydrothermal aging on the dielectric properties of epoxy-silica composites. The investigation was carried out with epoxy silica composites subjected to moisture exposure (standard Joint Electron Device Engineering Council [JEDEC] procedures) and consecutive thermal stress (240°C). They have stated that water absorption increases with curing.

They have also reported that after aging, the dielectric constant was found to be decreased and the loss factor was increased. In some specimens, polymer matrix composite (PMC) was performed at 165°C, for a duration of 1.5 h, 3 h, or 12 h. The cross-linking density increases to 93%, 98%, and 99% for curing times of 1.5, 3, and 12 h, respectively. The cured epoxy composites have a higher resistance against hydrothermal aging.

Figure 6.1 indicates the water absorption behavior resulting after hydrothermal conditioning. Water uptake was increased with the cross-linking density (or with PMC time). The same trend was also seen by Ko in similar composites [30]. The same was described in pure epoxy [31], which means that the effect is inherent to the epoxy matrix. Water diffusion is normally assumed to be lower in "structured" areas (highly cross-linked domains) than in "disordered" domains (loose or partially cured domains) [3,31]. Maggana et al. [3] also found that loose domains permit better segmental mobility, and so larger water mobility. Ko et al. [30] used the following statement to describe higher water absorption in the case of cured samples. According to them, by bridging lengthy molecular segments the cross-linking points check dense molecular packing. As a result, cross-linked areas have a higher free volume and can retain extra water content [30]. Higher water absorption involves both the presence of domains with a high free volume (water accommodation, probably in cross-linked areas) and the existence of an easy access path to these domains (water diffusion, possibly in loose areas). Epoxies are certainly known to be created of various phases, as confirmed by the anisotropic resistance against chemical attack [31]. Water absorption behavior is possibly to be determined by the way these phases are spatially disseminated, and by the way their distribution changes upon curing [31].

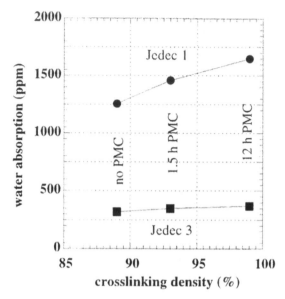

FIGURE 6.1
Water uptake after moisture exposure (Jedec 1: 85% RH, 85°C, 168 h; Jedec 3: 60% RH, 30°C, 192 h) and thermal stress (three consecutive rises to 240°C), as a function of cross-linking density (or PMC time). (From Gonon, P. et al., *Mater. Sci. Eng. B.*, 83, 158–164, 2001.)

Liu et al. [32] studied the effect of SiO_2 nanoparticles on the hydrophobic properties of waterborne fluorine-containing epoxy coatings. They have reported that the presence of SiO_2 significantly improved the corrosion resistance of the composite coatings, i.e., two times more than that of the neat waterborne fluorine-containing epoxy coatings. When the nanosilica content is 2.0%, the coating has the largest contact angle.

Han et al. [33] investigated the fracture toughness and wear properties of nanosilica/epoxy composites under a marine environment. They have used an epoxy-based nano-composite with 2 wt% nano-SiO_2 particles to increase the fracture toughness and wear performance of the neat epoxy under a marine environment. The mechanical properties of the neat epoxy and the nano-SiO_2-filled epoxy composite are categorized after various time durations of immersion in seawater (0 day, 7 days, and 30 days). The experimental effects show that immersion in seawater does not significantly affect the tensile behaviors of the epoxy or the nano-SiO_2-filled epoxy composite. However, the fracture toughness property of the neat epoxy was decreased by the immersion in seawater with an 18.5% reduction after 7 days of immersion and no further decrease was observed after 30 days of immersion.

Figure 6.2 reveals the dynamic mechanical properties of the neat epoxy polymer and nanosilica epoxy polymer nano-composites. Figure 6.2 shows

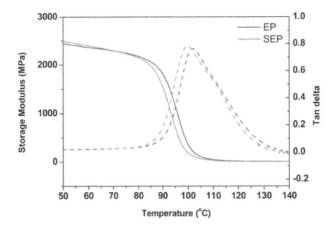

FIGURE 6.2
Storage modulus and loss tangent of neat epoxy (EP) and nanosilica-enhanced epoxy composite (SEP) as a function of temperature. (From Han, W. et al., *Mater. Chem. Phys.*, 177, 147–155, 2016.)

the storage modulus and tangent delta as a function of temperature. The T_g of a sample can be measured at the temperature where the tan delta rises up to its maximum value. From Figure 6.2, it can be observed that with the incorporation of nano-SiO_2, the T_g values are very close with 102°C for the epoxy and 100°C for the nano-SiO_2-enhanced composite, suggesting that the nano-SiO_2-enhanced composites have similar network features as that of the epoxy [34]. The dynamic mechanical properties of the epoxy polymer were slightly affected by the incorporation of nano-SiO_2 and temperature, the storage modulus of the nano-composite has a slight higher value than that of the epoxy at the same temperature, for example, at 50°C, the storage modulus of the epoxy is 2440 MPa, while the nano-SiO_2-enhanced composite has a value of 2490 MPa, which is steady with the improvement in stiffness value. The storage modulus of both specimens reduced significantly until coming to their respective T_g and have the similar values when the temperature was higher than T_g. So, with the small addition of nano-SiO_2 particles, the dynamic mechanical properties of the epoxy polymer have not been affected significantly.

Typical SEM micrographs of the fracture surfaces of the neat and the modified epoxies after saltwater immersion are shown in Figure 6.3. The fracture surface of the neat epoxy near the crack tip is glossy and brittle, as shown in Figure 6.3a. The fracture surfaces of the composite manifest small crack trajectories that are reflected and meander through the matrix, giving rise to the increased resistance to crack propagation, as shown in Figure 6.3b. After saltwater immersion (Figure 6.3c and d), many microcracks were generated along the crack initiation site. Meanwhile, comparing the neat epoxy without saltwater immersion (Figure 6.3a) and after immersion (Figure 6.3c and e), the density of crack lines decreases in the fracture surfaces. According to

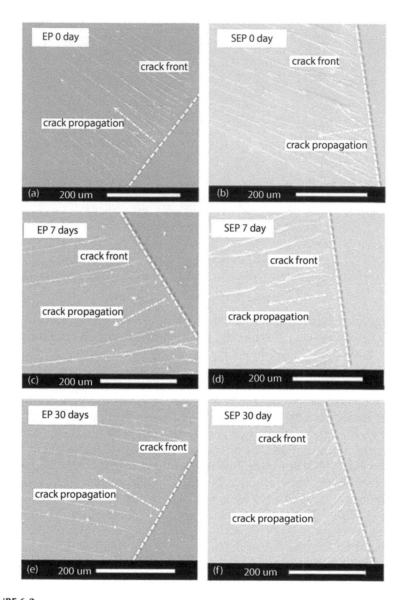

FIGURE 6.3
SEM images of fracture surfaces of (a), (c) and (e) neat epoxy (EP), and (b), (d), and (f) nanosilica-reinforced epoxy composite (SEP) after (a) and (b) 0 days, (c) and (d) 7 days, and (e) and (f) 30 days of salt water immersion. (From Han, W. et al., *Mater. Chem. Phys.*, 177, 147–155, 2016.)

previous research [34], the addition of a small amount of nanoparticles into the epoxy coating significantly reduces the rate of corrosion. The SiO_2 nanoparticles tend to occupy interspacing in the resin and serve to bridge more molecules in the interconnected matrix, leading to improved corrosion protection for the inner matrix.

6.3 Al₂O₃

Zhao et al. [28] studied the effect of water absorption behavior on the mechanical and dielectric properties of the Al_2O_3/epoxy nano-composites. The results obtained from nano-Al_2O_3-embedded epoxy composites show improvement in the tensile properties. The dynamic mechanical analysis (DMA) results reveal the enhancement in the stiffness value at the rubbery state of the matrix. The mechanical properties of the Al_2O_3/epoxy nanocomposites decrease with water absorption. This may be attributed to the damage and degradation of water on the epoxy matrix. However, the ductility of the nano-Al_2O_3/epoxy can be enhanced by the water absorption method. Results obtained from the water absorption behavior are shown in Figure 6.4, and it can be seen that the nature of water absorption curves for the epoxy and the nano-Al_2O_3/epoxy are almost the same to each other. This means that the incorporation of nano-Al_2O_3 particles into the epoxy matrix with various concentrations do not change the basic water diffusion mechanism. At any particular water treatment time (revealed as $t^{1/2}/h$ in Figure 6.4), the percentage of water absorption has been decreased with increasing the nano-Al_2O_3 content. This may be attributed to two features. Firstly, the volume of epoxy for water diffusion is reduced with increasing the nano-Al_2O_3 concentration. Secondly, the existence of the nano-Al_2O_3 can enhance the path length for moisture diffusion [35]. Usually, spherical nano-Al_2O_3 particles are not as active as layer silicates in introducing torturous-paths for water molecules to diffuse [36].

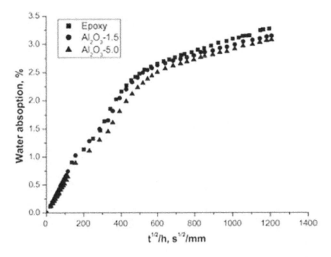

FIGURE 6.4
Curves of water absorption versus the square root time per specimen thickness, $t^{1/2}/h$ ($s^{1/2}$ mm), for epoxy and selected alumina-filled nano-composites. (From Zhao, H. and Li, R.K.Y., *Compos. Part Appl. Sci. Manuf.*, 39, 602–611, 2008.)

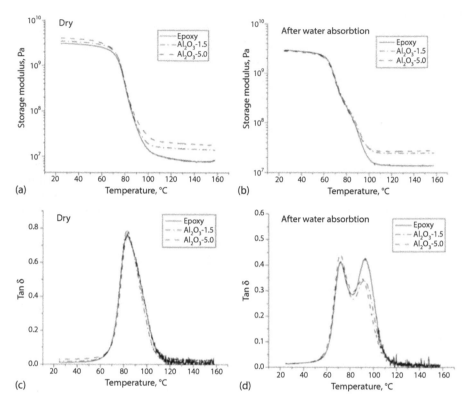

FIGURE 6.5
DMA results for epoxy, Al_2O_3-1.5, and Al_2O_3-5.0 nano-composites: (a) storage modulus for the dry samples; (b) storage modulus for the water-treated samples; (c) tanδ for the dry samples, and (d) tanδ for the water-treated samples. (From Zhao, H. and Li, R.K.Y., *Compos. Part Appl. Sci. Manuf.*, 39, 602–611, 2008.)

Figure 6.5a–d shows the DMA effects for the epoxy and the nano-Al_2O_3-embedded epoxy matrix nano-composites containing 1.5 and 5.0 phr. Figure 6.5a and b depicts the storage modulus (E) vs. temperature curves for the dry- and water-conditioned samples, respectively. When comparing both figures, it was evident that at temperatures below T_g, there is an apparent variance about the effect of the nano-Al_2O_3 content on E'. In the case of the dry specimens, E' increases marginally with increasing nano-Al_2O_3 content; while the nano-Al_2O_3 content has no noticeable effect on E' for the water-conditioned specimens. However, at temperatures higher than T_g, i.e., when the epoxy polymer is in the rubbery state, there are clear enhancements in E' with increasing the nano-Al_2O_3 content for both the dry- and water-conditioned nano-composites. The modification in reinforcement efficiency of the nano-Al_2O_3 particles in the epoxy matrix at temperatures below and above T_g can be described as follows. Rigid nano-Al_2O_3 particles can act as physical cross-links for the epoxy molecular chains in the nano-composites [37].

At temperatures higher than the T_g of the epoxy, the soft rubbery nature of the matrix causes the stiffness to decrease dramatically. In this case, the stiffening result of the nano-Al_2O_3 particles becomes more imperative. As a result, the magnitude of E' increases more significantly with the nanoparticle content. The tanδ vs. temperature plots for the nano-composites in the dry- and water-conditioned samples are revealed in Figure 6.5c and d, respectively. In both the dry- and water-conditioned states, it can be clearly seen that the T_g of the epoxy was not much affected by the incorporation of the nano-Al_2O_3 particles. The T_g values for the water-treated samples were lowered by 11°C as compared to the dry specimens. This indicates a decrease in the T_g value for water-absorbed specimens. This may be attributed to the plasticization effect of the water molecules diffused into the epoxy matrix. These diffused water molecules formed hydrogen bonds with the hydroxyl group of the epoxy molecular chain [3,38].

Figure 6.5d shows an additional tanδ peak in the epoxy and the nano-Al_2O_3 composites after water absorption. The presence of two peaks illustrates that the material structure is heterogeneous, which is directly initiated by water diffusion. The water molecules that were not chemically bonded onto the epoxy molecular chains will evaporate above 100°C and dissipate energies, affecting the development of the second peak [6]. As to the effect of the nano-Al_2O_3 on the tanδ curves for the epoxy, it can be seen that the existence of nano-Al_2O_3 particles does not alter the damping property of the epoxy, since the shape and position of the tanδ peaks for the nano-composites are the same as that of the unmodified epoxy. Therefore, the nano-Al_2O_3 particles have little effect on the energy dissipation process of the epoxy matrix under the testing conditions [28].

Nayak [22] studied the impact of seawater aging on the mechanical properties of nano-Al_2O_3-embedded glass fiber reinforced polymeric (GFRP) nano-composites. The plain and nano-composites were seawater-aged in a temperature controlled seawater bath at 70°C for a duration of 40 days. The results showed that seawater-aged plain and nano-composites deteriorate their flexural strength, interlaminar shear strength, and glass transition temperature. However, with the addition of 0.1 wt% of nano-Al_2O_3 particles into the plain GFRP composite, the flexural strength increased by 12%, interlaminar shear strength by 11% in the dry condition, and the seawater diffusion coefficient was reduced by 17% as compared to the plain GFRP composites. Nevertheless, the T_g of the nano-composites was not improved in dry and seawater-aged conditions significantly.

Figure 6.6a illustrates seawater gain (wt%) as a function of the square root of time. It was observed that with the increase in the nano-Al_2O_3 content in the GFRP composites, the water gain wt% increases. Nano-Al_2O_3 particles improve the interfacial bonding between fiber and matrix, and meanwhile, it increases the void percentage in the nano-enhanced GFRP composites. As a result, the decrease of water absorption with the increase in nano-Al_2O_3 content is not realized. Figure 6.6b depicts the effect of the nano-Al_2O_3 content on the diffusion coefficient of seawater. It was experientially

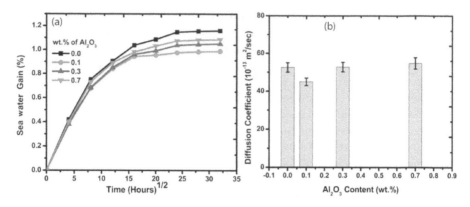

FIGURE 6.6
(a) Absorbed seawater wt% as a function of the square root of aging time and (b) diffusion coefficient versus nano-Al_2O_3 content (wt%) of the composites. (From Nayak, R.K., *Constr. Build. Mater.*, 221, 12–19, 2009.)

seen that the nano-composites having 0.1 wt% nano-Al_2O_3 have decreased the diffusion coefficient by 17% as compared to the neat GFRP composites. This may be attributed to the variation in the thermal expansion of the epoxy (6.2×10^{-5} K^{-1}) [39], fiber ($5–12 \times 10^{-6}$ K^{-1}) [40], and nano-Al_2O_3 particles (8.1×10^{-6} K^{-1}) [41] developing matrix swelling and microcrack formation at the interface. Hence, it was estimated that with the increase in moisture content of the nano-enhanced GFRP composites, the mechanical and thermal behavior would be degraded [42,43].

Figure 6.7 reveals the effect of the nano-Al_2O_3 content on: (a) flexural strength, (b) flexural strain, and (c) modulus of dry and seawater-aged GFRP composites. The results illustrate that with the incorporation of 0.1 wt% of nano-Al_2O_3 into the epoxy polymer, the flexural strength enhanced by 11% in the dry state. Whereas in the seawater-aged condition specimens, the flexural strength of the 0.1 wt% of nano-Al_2O_3 content nano-composites was increased by 12% as compared to the neat GFRP composites. With the enhancement in the nano-Al_2O_3 content, flexural strain decreased, resulting in flexural modulus increases in the nano-composites [44]. The possible reason may be attributed to the degradation of the epoxy polymer in seawater enhanced its brittleness of the nano-enhanced GFRP composites. The improvement of the mechanical properties may be attributed to a better interface bond between the epoxy and glass fiber at 0.1 wt% nano-Al_2O_3 content, resulting in the reduction of water diffusion into the nano-composites. However, with further enhancement in the nano-Al_2O_3 content, a reduction in the flexural strength was observed. This may be due to the increase in the nano-Al_2O_3 content, agglomeration, and void content of the nano-Al_2O_3 particle, leading to matrix swelling and microcrack formation at the interface region [44].

Nayak et al. [45] investigated the water absorption and thermal and mechanical behavior of hydrothermally exposed nano-Al_2O_3-filled glass

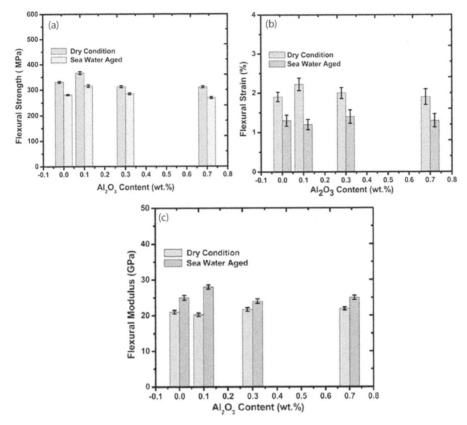

FIGURE 6.7
Comparison of (a) flexural strength, (b) flexural strain, and (c) modulus versus nano-Al_2O_3 content in dry- and seawater-aged condition. (From Nayak, R.K., *Constr. Build. Mater.*, 221, 12–19, 2019.)

FRP composites. The durability and reliability of the nano-Al_2O_3-filled GFRP composites in a hydrothermal condition is essential for hydro/hygro thermal applications. The study highlights the effect of the nano-Al_2O_3 particles concentration on the moisture absorption kinetics, residual mechanical, and thermal properties of hydrothermally conditioned GFRP nano-composites. Nano-Al_2O_3 particles were thoroughly mixed with the epoxy polymer through a temperature-assisted magnetic stirrer and followed by ultrasonication treatment. The results revealed that the incorporation of 0.1 wt% of nano-Al_2O_3 into the GFRP composites decreases the moisture diffusion coefficient by 10%, and in the meantime improves the flexural residual strength by 16% and interlaminar residual shear strength by 17%, as compared to the neat epoxy GFRP composites. However, the glass transition temperature has not been improved by the addition of the nano-Al_2O_3 filler, as compared to the neat GFRP composite.

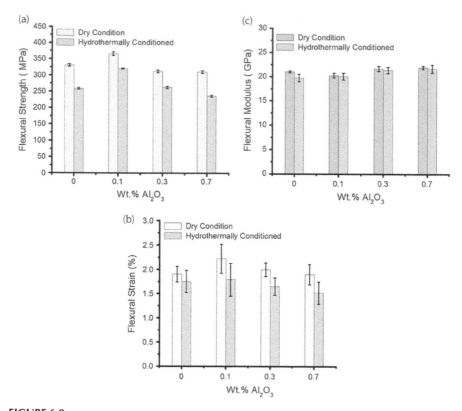

FIGURE 6.8
(a) Flexural strength, (b) strain, and (c) modulus of GFRP composites in dry and hydrothermal conditions with different nano-Al$_2$O$_3$ content. (From Nayak, R.K. and Ray, B.C., *Polym. Bull.*, 74, 4175–4194, 2014.)

Figure 6.8 reveals the effect of the nano-Al$_2$O$_3$ content on flexural strength, strain, and modulus of GFRP composites. It was observed that the incorporation of 0.1 wt% of nano-Al$_2$O$_3$ to the epoxy matrix enhances the flexural strength of the GFRP composites. The further increase in wt% of nano-Al$_2$O$_3$ decreases the flexural strength of the composites. The reason may be attributed to the increase in wt% of the nano-Al$_2$O$_3$ particles, the higher formation of agglomeration, decrease of the interfacial bond, and formation of more numbers of isolated pores/voids in the GFRP composites. The abovementioned reasons lead to the formation of a microcrack, resulting in the decrease in the flexural strength of the composites. However, the flexural modulus improves with an increase in the nano-Al$_2$O$_3$ content. A similar observation was also stated by Li et al. [46].

Figure 6.9 reveals interlaminar shear strength (ILSS) with a variation in the nano-Al$_2$O$_3$ content for dry- and hydrothermally conditioned samples. It was observed that the ILSS value has been improved for both dry- and

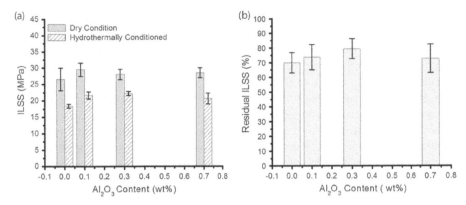

FIGURE 6.9
(a) Interlaminar shear strength in dry and hydrothermally conditioned samples (b) residual interlaminar shear strength versus nano-Al_2O_3 content (wt%). (From Nayak, R.K. and Ray, B.C., *Polym. Bull.*, 74, 4175–4194, 2017.)

hydrothermally conditioned specimens in nano-enhanced GFRP composites, as compared to the control GFRP composites. The incorporation of 0.1 wt% of nano-Al_2O_3 in the epoxy matrix enhances 17% of residual ILSS, as compared to the control GFRP composites. The enhancement of the ILSS value is attributed to the improvement of the interface strength, as compared to the reduction in micropores. The decrease in the ILSS value with an increase in the nano-Al_2O_3 content is attributed to the agglomeration of nano-Al_2O_3 particles, which shrink the effective interface surface area, resulting in the decrease of the interface shear strength. The reduction in the ILSS value of hydrothermally treated specimens is attributed to the formation of microcracks in the polymer matrix and interphase and interfacial debonding. The formation of microcracks at the interphase region is due to the differential expansion of the matrix and fiber. This is because of the water absorption and difference in the thermal gradient during hydrothermal conditioning [9–11].

6.4 TiO_2

This section highlights the incorporation of nano-TiO_2 fillers into the polymer matrix on the absorption of water and mechanical and thermal behavior of FRP composites. Nayak [47] studied the effect of nano-TiO_2 fillers on the mechanical behavior of hydrothermal-aged glass FRP composites. They have investigated the addition of different amounts of nanofillers content in the epoxy matrix and evaluated the water absorption and mechanical and thermal behavior of glass FRP composites.

Nayak et al. [44] investigated the influence of seawater absorption on the retention of mechanical properties of the nano-TiO_2-embedded glass fiber reinforced epoxy polymer matrix composites. The nano-TiO_2 particles are incorporated with the epoxy polymer matrix of GFRP composites to increase the mechanical properties. In the research exploration, the nano-TiO_2 particles of several weight fractions were thoroughly mixed with the epoxy polymer to increase the mechanical properties. The nano-enhanced GFRP composites are fabricated by using the hand lay-up technique and subsequently immersed in a seawater bath at 70°C for 40 days. The seawater diffusivity, flexural, and interlaminar shear strength of the neat GFRP and nano-TiO_2-enhanced GFRP composites were calculated and compared between themselves. The result indicated that with the incorporation of 0.1 wt% of the nano-TiO_2 fillers into the epoxy polymer, the seawater diffusivity enhanced by 15%. However, the flexural and ILSS of seawater-conditioned nano-TiO_2-enhanced GFRP composites were increased by 15% and 23%, respectively.

Figure 6.10a reveals the seawater gain wt% with the square root of the seawater-aging duration. The result revealed that the water gain percentage increases with time in the initial stage of aging and, later on, it decreases with the further increase in time. It may be due to the presence of the open pores or voids on the surface of the composite that helps to accelerate the absorption tendency. However, the water absorption increases with the increase in the nano-TiO_2 content. The seawater absorption reduced for the nano-composites having 0.1 wt% of the nano-TiO_2 particles. It may be due to a good interface bond between the matrix/fiber, which reduces the water diffusion into the composites through capillary action. With a further increase in wt% of the nano-TiO_2 content, void content (%) becomes the dominating factor to absorb moisture in the composites. Figure 6.10b shows the seawater diffusion coefficient versus wt% of the nano-TiO_2 in the nano-composites. The linear portion of the absorbed water wt% versus the square root of the

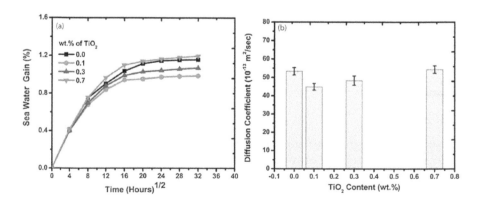

FIGURE 6.10
(a) Seawater gain (%) versus the square root of time (b) diffusivity versus nano-TiO_2 content. (From Nayak, R.K. and Ray, B.C., *Arch. Civ. Mech. Eng.*, 18, 1597–1607, 2018.)

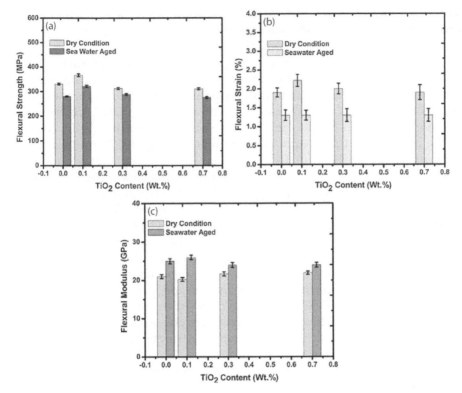

FIGURE 6.11
The wt% of nano-TiO$_2$ content versus flexural strength (a), strain (b), and modulus (c) of the composites. (From Nayak, R.K. and Ray, B.C., *Arch. Civ. Mech. Eng.*, 18, 1597–1607, 2018.)

time plot is assumed to be Fickian diffusion. The result revealed that the diffusion coefficient increases with the increase in the nano-TiO$_2$ content in the nano-composite laminates. However, with the addition of 0.1 wt% of nano-TiO$_2$ into the epoxy matrix, the seawater diffusivity reduces by 15%.

Figure 6.11 shows: (a) flexural strength, (b) strain, and (c) modulus versus wt% of nano-TiO$_2$ in the composites. The results revealed that with the increase in wt% of the nano-TiO$_2$ content, the flexural strength reduces. The maximum improvement of the flexural strength was observed for the nano-composites having 0.1 wt% of nano-TiO$_2$, around 11%, i.e., from 330 to 366 MPa. However, it reduces with the further increase in wt% of nano-TiO$_2$ in the composites. The improvement of the flexural strength after seawater aging is around 15% for 0.1 wt% of the nano-TiO$_2$ content composite. The improvement of the flexural strength is attributed to a better interface/interphase bond between the fiber and matrix [16]. Furthermore, with the increase in the nano-TiO$_2$ content, the van der Waals' force between two nanoparticles increases the resulting agglomeration of nanoparticles in the epoxy matrix. The agglomerated nanoparticles reduce the active surface area to interact with the epoxy matrix, resulting in ineffective load transfer at the fiber/matrix interface. Void content in the composites is also one of the reasons for the reduction of the flexural strength.

The decrease in the mechanical properties of the hydrothermally aged composites is because of uneven thermal expansion of the fiber, epoxy, and nanoparticles, which leads to interface swelling and hydrolysis of the epoxy polymer. It results in the degradation of interface/interphase bond between the matrix and fiber [8,9,31]. Hence, with increases in the nano-TiO_2 content in the composites, the flexural strength reduces. Figure 6.11b shows strain versus wt% nano-TiO_2. Initially, flexural strain increases with the addition of nano-TiO_2 particles and decreases with the further increases in the TiO_2 content in a dry condition. Figure 6.11c shows the flexural modulus versus the nano-TiO_2 content in the composites. In a dry condition, the flexural modulus increases with the increase in the nano-TiO_2 content. It may be due to the presence of a high modulus of the nano-TiO_2 particles in the composites. However, the flexural modulus of nano-composites reduces with the increase in the nano-TiO_2 content of seawater-aged composites.

Ng et al. [35] investigated the mechanical and permeability properties of nano- and micron-TiO_2-filled epoxy composites. Swelling investigations were accomplished by the introduction of 1 cm cubes of neat and filled epoxies in boiling water and measuring both the weight gain and the volume change with the time duration of 7 days.

A substantial constraint in several epoxies polymer is their attraction for water and the swelling that arises upon conditioning to water environments. To identify this behavior properly, the specimens were conditioned to harsh conditions of boiling water for numerous days. Figure 6.12 illustrates the change in weight for the unfilled, nano-filled, and micron-filled epoxies. Whereas the volume change for the micron-filled epoxy seems to enhance very slowly at the beginning, its volume change after 5 days is equal to that of the unfilled epoxy.

In contrast, the nano-enhanced composite showed much improved dimensional stability, presenting saturation in the volume change after only two days at a value only half that of the other two specimens. Although a minor modification in volume for the nano-enhanced epoxy could show the existence of an interaction region with decreased water absorption, the lack of a decrease in the weight gain of the nano-enhanced epoxy does not support this type of behavior. The curing agent may absorb the water molecules, but not swell, and thus the increase in weight would be the same as that of the unfilled epoxy, but the swelling would be decreased. The water permeability (calculated as a weight change) of the micron-filled epoxy, on the other hand (Figure 6.13), is lower than that for either the nano-filled or unfilled epoxies. A minute investigation of the swelling statistics illustrates that the micron-filled epoxy swells more slowly than the nano-filled epoxy, which reveals the lower transport rate phenomenon. Permeability often reduces in a filled polymer due to the increase in tortuosity, since

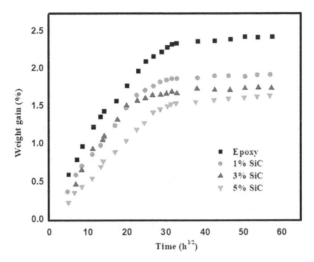

FIGURE 6.12
The results of exposing unfilled epoxy, 10 wt% nano-TiO$_2$-filled epoxy and 10 wt% micron TiO$_2$-filled epoxy composites to boiling water with respect to weight change. (From Ng, C.B. et al., *Adv. Compos. Lett.*, 10, 2001.)

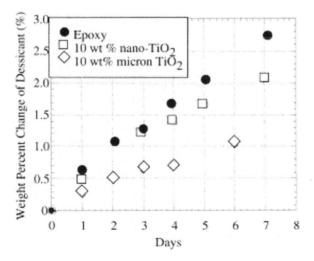

FIGURE 6.13
The permeability results for unfilled epoxy, 10 wt% nano-TiO$_2$-filled epoxy, and 10 wt% micron TiO$_2$-filled epoxy composites. (From Ng, C.B. et al., *Adv. Compos. Lett.*, 10, 2001.)

the path that the water must take in infiltrating a film is more complex due to the existence of the nanofillers [48]. In the case of the micron-sized filler, the path may be even further tortuous with the variation in dimensions of the filler particles, whereas the nano-sized particles do not alter the path significantly.

6.5 SiC

Alamri et al. [49] studied the effect of water absorption properties on the mechanical behavior of nanofiller-reinforced epoxy nano-composites as well as investigated the effect of various kinds of nanofillers such as halloysite nanotubes, nanoclay platelets, and nanosilicon carbide (n-SiC) particles on the water absorption properties of the epoxy polymer nano-composites. Both water uptake and diffusivity are found to be reduced with the incorporation of nanofillers into the epoxy matrix as compared to unfilled epoxy. It was observed that both flexural strength and the modulus of the epoxy nano-composites showed a reduction due to the water absorption. However, the incorporation of the nanofillers increased the flexural strength and modulus of the nano-composites compared to the wet unfilled epoxy. Furthermore, the impact strength and fracture toughness of all kinds of nano-filled composites were found to increase after conditioning to the water environment. The existence of nanoparticles improved both impact strength and fracture toughness of the nano-composites as compared to the wet neat epoxy composites.

Excitingly, the nano-composites enhanced with n-SiC indicate improved barrier properties as compared to other filled nano-composites. The addition of 1, 3, and 5 wt% n-SiC reduces the water uptake by 21.8%, 28.6%, and 33.3%, respectively, as compared to that of neat epoxy.

Alamri et al. [50] investigated the effect of the incorporation of n-SiC particles on the water absorption behavior of n-SiC-reinforced recycled cellulose fiber (RCF)/epoxy eco-nano-composites. Figure 6.14 shows the typical flexural stress-strain curves for unfilled RCF/epoxy and 5 wt% n-SiC-filled RCF/epoxy composites before and after placing in water. With the increase

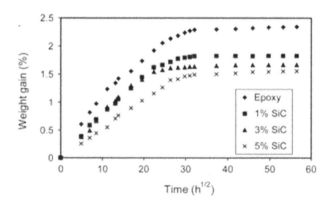

FIGURE 6.14
Water absorption curves of epoxy-based nano-composites filled with n-SiC. (From Alamri, H. and Low, I.M. et al., *Mater. Des.*, 42, 214–222, 2012.)

in percentage of n-SiC in composites, the water uptake decreased. The influence of the water absorption on the mechanical behavior of n-SiC-filled RCF/epoxy eco-nano-composites was studied as a function of the concentration of n-SiC. The water absorption in the polymer phase resulted in a significant decrease in the flexural strength, flexural modulus, and fracture toughness of n-SiC-enhanced RCF/epoxy nano-composites as compared to dry composites. The flexural strength, flexural modulus, and fracture toughness improved by 14.4%, 7.5%, and 6.1%, respectively, compared to unfilled RCF/epoxy composites after the incorporation of 5 wt% n-SiC. SEM results revealed a clean pullout of cellulose fibers as a result of the degradation in the fiber-matrix interfacial bonding by the water absorption. The incorporation of n-SiC was found to increase the interfacial adhesion between the matrix and fiber.

Figure 6.15 illustrates the influence of the incorporation of n-SiC particles on the water absorption behavior of the RCF/epoxy eco-nano-composites. The results showed that all composites possess a similar nature of curves. This indicates that the incorporation of n-SiC has no significant effect on the mechanism of the water absorption in RCF/epoxy composites. For all composites, the water uptake continuously enhances with the increasing time of immersion. However, the water absorption behavior is found to decrease gradually due to the existence of nanoparticles. It was stated in literature that the existence of a high aspect ratio nanofiller in a polymer matrix improves the barrier resistance properties of the materials by producing tortuous pathways for the water molecules to diffuse into the composites, which leads to a decrease in the absorbed water content [28,51]. The maximum water uptake of RCF/epoxy composites that are incorporated with 5 wt% n-SiC reduces by 47.5%, as compared to unfilled RCF/epoxy composites. Similarly, Mohan et al. [52] reported a dramatic decrease in the water mass uptake of nanoclay-filled sisal fiber/epoxy composites.

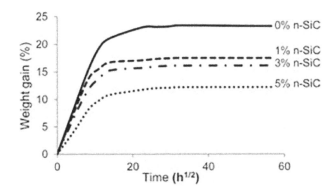

FIGURE 6.15
Water absorption curves of n-SiC-filled RCF/epoxy eco-nano-composites. (From Alamri, H. and Low, I.M. et al., *Polym. Test.*, 31, 810–818, 2012.)

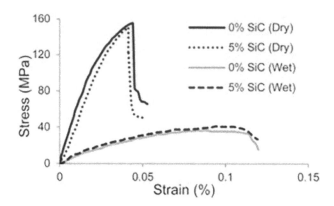

FIGURE 6.16
Typical stress-strain curves of RCF/epoxy composites filled with 0 and 5 wt% n-SiC in dry and wet conditions. (From Alamri, H. and Low, I.M., *Polym. Test.*, 31, 810–818, 2012.)

Figure 6.16 elucidates the flexural stress-strain curves for the neat RCF/epoxy and 5 wt% n-SiC-enhanced RCF/epoxy composites before and after immersing in the water. It was observed that the incorporation of n-SiC particles decreases the maximum stress of the RCF/epoxy composites in dry conditions. However, the existence of n-SiC leads to the enhanced maximum stress after water conditioning. It was also evident from Figure 6.16 that the flexural properties for both varieties of composites severely decrease due to the water absorption. The significant decrease in the maximum stress after the water conditioning can be attributed to the fiber damage and degradation in the fiber-matrix interfacial bonding due to the absorbed water [53]. Due to the water absorption, the maximum bending strain of the composites has been increased. The performance of the water molecules is like a plasticizer agent in FRP composites, leading to an increase in materials ductility, which is attributed to the increase in maximum strain [28,53,54].

6.6 CaCO₃

Eskizeybek et al. [55] studied the static and dynamic mechanical behaviors of $CaCO_3$-modified epoxy/carbon fiber nano-composites. The authors incorporated different amounts of nano-$CaCO_3$ into the epoxy polymer, and the nano-reinforced epoxy was used to impregnate carbon fabrics (CF) by utilizing the vacuum-assisted resin infusion method. The prepared FRP nano-composites were exposed to tensile, bending, and low velocity impact loadings. From the experimental result, it was observed that the tensile strength of the CF/epoxy nano-composites improved by 48% with the incorporation of 2 wt% $CaCO_3$ nanoparticles. In the same way, the increase in the flexural strength was also observed to be about 47% for the same content of nano-$CaCO_3$. Low-velocity impact tests were carried out, and the nano-$CaCO_3$-reinforced CF/epoxy composites exhibited higher impact performances as compared to the neat CF/epoxy composites. The fracture morphologies were observed by electron microscopy to reveal some toughening mechanisms. Debonding, crack deflection, and crack pinning of the nanoparticles were the primary causes prominent to the enhancement of toughness. The authors reported that the incorporation of the nano-$CaCO_3$ in the CF/epoxy composites has substantially altered the physical and mechanical properties of the nano-composites.

Kumar et al. [56] investigated the effects of fly ash and calcium carbonate fillers on the mechanical and moisture absorption properties in poly vinyl chloride resin.

The figures from Table 6.1 showed that the composite material with 11.35% of fly ash and 53.53% of fly ash showed the least changes to immersion of the material in water. They presented a constant rise in weight when the values were taken at 24-hour-time intervals. Furthermore, they indicated the minimum variation toward moisture conditioning when compared to all other compositions. On a general scale, 22.36% of the fly ash composition had a low enhancement in weight after 72 h of immersion in the water.

TABLE 6.1

Calculation of Moisture Absorption at Accelerated Hygrothermal Test for 72 h

Contains	Before Dipping the Specimen (Dry Weight) in "Grams"	After Dipping the Specimens (Wet Weight) in "Grams"		
		24 h	48 h	72 h
Resin-100%	13.78	13.87	13.88	13.89
Fly ash-11.35%	14.10	14.18	14.19	14.20
Fly ash-22.36%	15.20	15.26	15.28	15.29
Fly ash-33.05%	14.17	14.33	14.36	14.37
Fly ash-43.44%	18.66	18.76	18.77	18.79
Fly ash-53.53%	21.11	21.29	21.31	21.33
$CaCO_3$-11.35%	13.29	13.34	13.35	13.36
$CaCO_3$-22.36%	14.93	15.01	15.04	15.06
$CaCO_3$-33.05%	18.95	19.03	19.04	19.05
$CaCO_3$-43.44%	16.11	16.17	16.18	16.20
$CaCO_3$-53.53%	17.71	17.76	17.77	17.78

Source: Kumar, P.N. et al., *Mater. Today Proc.*, 16, 1219–1225, 2019.

Whereas 33.05% of the fly ash composition exhibited an extreme change in weight over the 72 h, which is approximately 0.20 grams. A similar observation is seen in the $CaCO_3$ filler incorporated in the composite material. Similar to the fly ash, 11.35% of the $CaCO_3$-embedded composites as well as 53.53% of the $CaCO_3$-enhanced composites revealed a minimum change in weight. At the same time, it displayed more resistivity to the water absorption, which is substantial since it absorbed only 0.01 grams of moisture at each period of 24 h.

6.7 BN

BN nanotubes possess numerous motivating properties. In comparison to carbon nanotubes, which are having about a zero or a very narrow band gap, whereas the BN nanotubes have a band gap of about 6 eV irrespective of the diameter or chirality change. BN nanotubes are capable of deep UV light emitters appropriate for optoelectronics and laser [57]. BN nanotubes have Young's modulus of 1.2 TPa, with outstanding stiffness and bending flexibility along the axial direction with a tubular structure and strong sp^2 bonding [58].

Kim et al. [59] studied BN nanoparticles that were comprised of various surface treatments to make thermally conductive polymer composites by epoxy wetting. The measured storage modulus was also improved by the surface treatment because of the sufficient interface formed and contact between the enormous quantity of the filler and epoxy.

The mechanical properties were recorded by a dynamic mechanical thermal analyzer to ensure the degree to which these properties were enhanced after the surface modification of the BN particles. Because the usual operating temperatures for electronic packaging are around 150°C, the steadiness of the composites was observed up to this temperature. The storage modulus of the composite with a filler content of 70 wt% as a function of temperature is showed in Figure 6.17. It is evident that the storage modulus of the composite correlated with the extent of the filler surface modification, which is due to the mechanical reinforcement resulting from the strong interaction between the BN particles and the epoxy-terminated dimethylsiloxane (ETDS) matrix. The incorporation of hydroxide groups enhanced the polar interactions, and therefore, the cohesion between the BN filler and the ETDS matrix. However, the stress transmissibility was not effective when the same force was applied to the composite. The hydroxyl groups in the composites stopped treeing from efficiently propagating the stress, which helps in increasing the storage modulus. Figure 6.17 revealed that the modulus of the BN-403 composite was slightly higher than that of the BN composite.

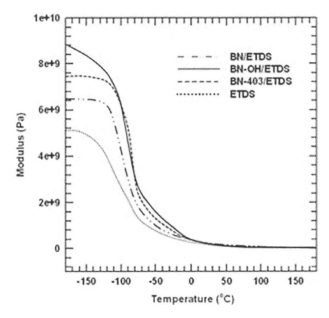

FIGURE 6.17

Storage modulus of ETDS and ETDS composites with 70 wt% filler concentration. (From Kim, K. and Kim, J., *Ceram. Int.*, 40, 5181–5189, 2014.)

6.8 Synergistic Effect of Nanofillers

Various researchers and investigators have found that individually or a combination of more than two nanofillers has positive responses on the mechanical properties of FRP composites [12]. However, the combined effect on the physical and mechanical properties of two or more than two nanoparticles at different weight percentages might be interesting and useful for the strength enhancement of composites.

Shi et al. [29] investigated the characterization of the protective performance of an epoxy reinforced with nanometer-sized TiO_2 and SiO_2. They have observed that the hardness of the epoxy resin was improved by the incorporation of 1 wt% nano-TiO_2 particles or nano-SiO_2. The results of salt spray tests evidenced the better protective performance of the epoxy coatings with the addition of nano-TiO_2 and treated nano-SiO_2 than the varnish and untreated nano-SiO_2 incorporated epoxy coatings. The comparison of the electrochemical impedance spectrum results show that the efficiency of the improvement of corrosion resistance for embedding nano-SiO_2 is higher than that of nano-TiO_2, which is indicated by the higher coating resistance in the low frequency part in the immersion period.

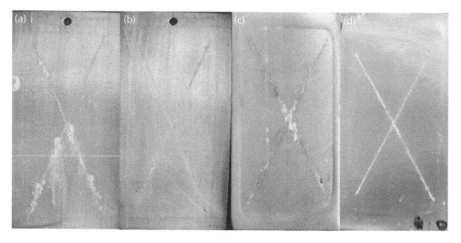

FIGURE 6.18
Aspects of (a) epoxy varnish, (b) varnish with nano-TiO$_2$, (c) varnish with untreated nano-SiO$_2$, and (d) varnish with treated nano-SiO$_2$, after exposure in salt spray for 500 h. (From Shi, H. et al., *Prog. Org. Coat.*, 62, 359–368, 2008.)

The corrosion resistance of the epoxy was estimated by the rusts and blistering along the "X" marks on 2024-T3 substrate. The aspects of the neat epoxy, nano-TiO$_2$-enhanced epoxy resin, nano-SiO$_2$ epoxy resin with untreated, and with treated nano-SiO$_2$, after exposure in salt spray for 500 h, are shown in Figure 6.18. After salt spray for 500 h, severe scorching was seen along the "X" marked for epoxy resin, with a diameter of scratches more than 5 mm. However, no apparent scratches along the marks were detected on the nano-TiO$_2$-filled epoxy surface. A similar situation occurred in the treated nano-SiO$_2$-filled epoxy resin surface. However, small scratches with a diameter of 0.2–1 mm could be perceived for the untreated nano-SiO$_2$-filled epoxy resin. The corrosion results coincided with the impedance spectra, implying that the incorporation of nanoparticles can efficiently avert the epoxy resin from scratching and delamination.

References

1. Alexandre, M., and P. Dubois. Polymer-layered silicate nanocomposites: Preparation, properties and uses of a new class of materials. *Materials Science and Engineering R: Reports* 28 (2000): 1–63. doi:10.1016/S0927-796X(00)00012-7.
2. Gonon, P., A. Sylvestre, J. Teysseyre, and C. Prior. Combined effects of humidity and thermal stress on the dielectric properties of epoxy-silica composites. *Materials Science and Engineering: B* 83 (2001): 158–164. doi:10.1016/S0921-5107(01)00521-9.

3. Maggana, C., and P. Pissis. Water sorption and diffusion studies in an epoxy resin system. *Journal of Polymer Science Part B: Polymer Physics* 37 (1999): 1165–1182.
4. Xiao, G. Z., M. Delamar, and M. E. R. Shanahan. Irreversible interactions between water and DGEBA/DDA epoxy resin during hygrothermal aging. *Journal of Applied Polymer Science* 65 (1997): 449–458.
5. Verghese, K. N. E., M. D. Hayes, K. Garcia, C. Carrier, J. Wood, J. R. Riffle, and J. J. Lesko. Influence of matrix chemistry on the short term, hydrothermal aging of vinyl ester matrix and composites under both isothermal and thermal spiking conditions. *Journal of Composite Materials* 33 (1999): 1918–1938. doi:10.1177/002199839903302004.
6. De'Nève, B., and M. E. R. Shanahan. Water absorption by an epoxy resin and its effect on the mechanical properties and infra-red spectra. *Polymer* 34 (1993): 5099–5105. doi:10.1016/0032-3861(93)90254-8.
7. Ray, B. C. Temperature effect during humid ageing on interfaces of glass and carbon fibers reinforced epoxy composites. *Journal of Colloid and Interface Science* 298 (2006): 111–117. doi:10.1016/j.jcis.2005.12.023.
8. Yilmaz, T., and T. Sinmazcelik. Effects of hydrothermal aging on glass–fiber/polyetherimide (PEI) composites. *Journal of Materials Science* 45 (2010): 399–404. doi:10.1007/s10853-009-3954-1.
9. Gautier, L., B. Mortaigne, and V. Bellenger. Interface damage study of hydrothermally aged glass-fibre-reinforced polyester composites. *Composites Science and Technology* 59 (1999): 2329–2337. doi:10.1016/S0266-3538(99)00085-8.
10. Hodzic, A., J. K. Kim, A. E. Lowe, and Z. H. Stachurski. The effects of water aging on the interphase region and interlaminar fracture toughness in polymer–glass composites. *Composites Science and Technology* 64 (2004): 2185–2195. doi:10.1016/j.compscitech.2004.03.011.
11. Ellyin, F., and R. Maser. Environmental effects on the mechanical properties of glass-fiber epoxy composite tubular specimens. *Composites Science and Technology* 64 (2004): 1863–1874. doi:10.1016/j.compscitech.2004.01.017.
12. Nayak, R. K., K. K. Mahato, B. C. Routara, and B. C. Ray. Evaluation of mechanical properties of Al_2O_3 and TiO_2 nano filled enhanced glass fiber reinforced polymer composites. *Journal of Applied Polymer Science* 133 (2016). doi:10.1002/app.44274.
13. Hu, Y., G. Du, and N. Chen. A novel approach for Al_2O_3/epoxy composites with high strength and thermal conductivity. *Composites Science and Technology* 124 (2016): 36–43. doi:10.1016/ j.compscitech.2016.01.010.
14. Nayak, R. K., K. K. Mahato, and B. C. Ray. Water absorption behavior, mechanical and thermal properties of nano TiO_2 enhanced glass fiber reinforced polymer composites. *Composites Part A: Applied Science and Manufacturing* 90 (2016): 736–747. doi:10.1016/j.compositesa.2016.09.003.
15. Sprenger, S. Epoxy resins modified with elastomers and surface-modified silica nanoparticles. *Polymer* 54 (2013): 4790–4797. doi:10.1016/j.polymer.2013.06.011.
16. Al-Turaif, H. A. Effect of nano TiO_2 particle size on mechanical properties of cured epoxy resin. *Progress in Organic Coatings* 69 (2010): 241–246. doi:10.1016/j.porgcoat.2010.05.011.
17. Sharifi Golru, S., M. M. Attar, and B. Ramezanzadeh. Studying the influence of nano-Al_2O_3 particles on the corrosion performance and hydrolytic degradation resistance of an epoxy/polyamide coating on AA-1050. *Progress in Organic Coatings* 77 (2014): 1391–1399. doi:10.1016/j.porgcoat.2014.04.017.

18. Liang, Y. L., and R. A. Pearson. Toughening mechanisms in epoxy–silica nanocomposites (ESNs). *Polymer* 50 (2009): 4895–4905. doi:10.1016/j.polymer.2009.08.014.
19. Landowski, M., G. Strugała, M. Budzik, and K. Imielińska. Impact damage in SiO₂ nanoparticle enhanced epoxy–Carbon fibre composites. *Composites Part B: Engineering* 113 (2017): 91–99. doi:10.1016/j.compositesb.2017.01.003.
20. Fan, X. J., S. W. R. Lee, and Q. Han. Experimental investigations and model study of moisture behaviors in polymeric materials. *Microelectronics Reliability* 49 (2009): 861–871. doi:10.1016/j.microrel.2009.03.006.
21. Prusty, R. K., D. K. Rathore, and B. C. Ray. Water-induced degradations in MWCNT embedded glass fiber/epoxy composites: An emphasis on aging temperature. *Journal of Applied Polymer Science* 135 (2018): 45987. doi:10.1002/app.45987.
22. Nayak, R. K. Influence of seawater aging on mechanical properties of nano-Al₂O₃ embedded glass fiber reinforced polymer nanocomposites. *Construction and Building Materials* 221 (2019): 12–19. doi:10.1016/j.conbuildmat.2019.06.043.
23. Zulfli, N. M., A. A. Bakar, and W. Chow. Mechanical and water absorption behaviors of carbon nanotube reinforced epoxy/glass fiber laminates. *Journal of Reinforced Plastics and Composites* 32 (2013): 1715–1721. doi:10.1177/0731684413501926.
24. Okhawilai, M., I. Dueramae, C. Jubsilp, and S. Rimdusit. Effects of high nano-SiO₂ contents on properties of epoxy-modified polybenzoxazine. *Polymer Composites* 38 (2017): 2261–2271. doi:10.1002/pc.23807.
25. Kinloch, A. J., K. Masania, A. C. Taylor, S. Sprenger, and D. Egan. The fracture of glass-fibre-reinforced epoxy composites using nanoparticle-modified matrices. *Journal of Materials Science* 43 (2008): 1151–1154. doi:10.1007/s10853-007-2390-3.
26. Bajpai, P. L. *Reinforced Polymer Composites.* https://www.bookdepository.com/Reinforced-Polymer-Composites-Pramendra-K-Bajpai/9783527345991 (Accessed August 29, 2019).
27. Domun, N., H. Hadavinia, T. Zhang, T. Sainsbury, G. H. Liaghat, and S. Vahid. Improving the fracture toughness and the strength of epoxy using nanomaterials—A review of the current status. *Nanoscale* 7 (2015): 10294–10329. doi:10.1039/C5NR01354B.
28. Zhao, H., and R. K. Y. Li. Effect of water absorption on the mechanical and dielectric properties of nano-alumina filled epoxy nanocomposites. *Composites Part A: Applied Science and Manufacturing* 39 (2008): 602–611. doi:10.1016/j.compositesa.2007.07.006.
29. Shi, H., F. Liu, L. Yang, and E. Han. Characterization of protective performance of epoxy reinforced with nanometer-sized TiO₂ and SiO₂. *Progress in Organic Coatings* 62 (2008): 359–368. doi:10.1016/j.porgcoat.2007.11.003.
30. Ko, M., and M. Kim. Effect of postmold curing on plastic IC package reliability. *Journal of Applied Polymer Science* 69 (1998): 2187–2193.
31. Diamant, Y., G. Marom, and L. J. Broutman. The effect of network structure on moisture absorption of epoxy resins. *Journal of Applied Polymer Science* 26 (1981): 3015–3025.
32. Liu, X., R. Wang, Y. Fan, and M. Wei. Effect of SiO₂ nanoparticles on the hydrophobic properties of waterborne fluorine-containing epoxy coatings. *MATEC Web of Conferences* 130 (2017): 08005. doi:10.1051/matecconf/201713008005.

33. Han, W., S. Chen, J. Campbell, X. Zhang, and Y. Tang. Fracture toughness and wear properties of nanosilica/epoxy composites under marine environment. *Materials Chemistry and Physics* 177 (2016): 147–155. doi:10.1016/j.matchemphys.2016.04.008.

34. Zeng, S., C. Reyes, J. Liu, P. A. Rodgers, S. H. Wentworth, and L. Sun. Facile hydroxylation of halloysite nanotubes for epoxy nanocomposite applications. *Polymer* 55 (2014): 6519–6528. doi:10.1016/j.polymer.2014.10.044.

35. Ng, C. B., B. J. Ash, L. S. Schadler, and R. W. Siegel. A study of the mechanical and permeability properties of nano- and micron-TiO$_2$ filled epoxy composites. *Advanced Composites Letters* 10 (2001). doi:10.1177/096369350101000301.

36. Bharadwaj, R. K. Modeling the barrier properties of polymer-layered silicate nanocomposites. *Macromolecules* 34 (2001): 9189–9192. doi:10.1021/ma010780b.

37. Sue, H.-J., K. T. Gam, N. Bestaoui, N. Spurr, and A. Clearfield. Epoxy nano-composites based on the synthetic α-zirconium phosphate layer structure. *Chemistry of Materials* 16 (2004): 242–249. doi:10.1021/cm030441s.

38. Moy, P., and F. E. Karasz. Epoxy-water interactions. *Polymer Engineering & Science* 20 (1980): 315–319. doi:10.1002/pen.760200417.

39. Huang, C. J., S. Y. Fu, Y. H. Zhang, B. Lauke, L. F. Li, and L. Ye. Cryogenic proper-ties of SiO$_2$/epoxy nanocomposites. *Cryogenics* 45 (2005): 450–454. doi:10.1016/j.cryogenics.2005.03.003.

40. Chu, X. X., Z. X. Wu, R. J. Huang, Y. Zhou, and L. F. Li. Mechanical and thermal expansion properties of glass fibers reinforced PEEK composites at cryogenic temperatures. *Cryogenics* 50 (2010): 84–88. doi:10.1016/j.cryogenics.2009.12.003.

41. Corundum, Aluminum Oxide, Alumina, 99.9%, Al$_2$O$_3$, (n.d.). http://www.matweb.com/search/DataSheet.aspx?MatGUID=c8c56ad547ae4cfabad15977bfb537f1 (Accessed September 8, 2019).

42. Pervin, F., Y. Zhou, V. K. Rangari, and S. Jeelani. Testing and evaluation on the thermal and mechanical properties of carbon nano fiber reinforced SC-15 epoxy. *Polymer Engineering & Science* 405 (2005): 246–253. doi:10.1016/j.msea.2005.06.012.

43. Tritt, T. M., and D. Weston. Measurement techniques and considerations for determining thermal conductivity of bulk materials. In T. M. Tritt (Ed.), *Thermal Conductivity*, Springer, Boston, MA, 2004: pp. 187–203. doi:10.1007/0-387-26017-X_8.

44. Nayak, R. K., and B. C. Ray. Influence of seawater absorption on retention of mechanical properties of nano-TiO$_2$ embedded glass fiber reinforced epoxy polymer matrix composites. *Archives of Civil and Mechanical Engineering* 18 (2018): 1597–1607. doi:10.1016/j.acme.2018.07.002.

45. Nayak, R. K., and B. C. Ray. Water absorption, residual mechanical and ther-mal properties of hydrothermally conditioned nano-Al$_2$O$_3$ enhanced glass fiber reinforced polymer composites. *Polymer Bulletin* 74 (2017): 4175–4194. doi:10.1007/s00289-017-1954-x.

46. Li, W., A. Dichiara, J. Zha, Z. Su, and J. Bai. On improvement of mechanical and thermo-mechanical properties of glass fabric/epoxy composites by incorpo-rating CNT–Al$_2$O$_3$ hybrids. *Composites Science and Technology* 103 (2014): 36–43. doi:10.1016/j.compscitech.2014.08.016.

47. Nayak, R. K. Effect of nano-TiO$_2$ particles on mechanical properties of hydrothermal aged glass fiber reinforced polymer composites. In R. Prasad, T. Karchiyappan (Eds.), *Advanced Research in Nanosciences for Water Technology*, Springer International Publishing, Cham, Switzerland, 2019: pp. 69–93. doi:10.1007/978-3-030-02381-2_4.

48. Nielsen, L. E. Models for the permeability of filled polymer systems. *Journal of Macromolecular Science: Part A-Chemistry* 1 (1967): 929–942. doi:10.1080/10601326708053745.
49. Alamri, H., and I. M. Low. Effect of water absorption on the mechanical properties of nano-filler reinforced epoxy nanocomposites. *Materials & Design* 42 (2012): 214–222. doi:10.1016/j.matdes.2012.05.060.
50. Alamri, H., and I. M. Low. Effect of water absorption on the mechanical properties of n-SiC filled recycled cellulose fibre reinforced epoxy eco-nanocomposites. *Polymer Testing* 31 (2012): 810–818. doi:10.1016/j.polymertesting.2012.06.001.
51. Liu, W., S. V. Hoa, and M. Pugh. Fracture toughness and water uptake of high-performance epoxy/nanoclay nanocomposites. *Composites Science and Technology* 65 (2005): 2364–2373. doi:10.1016/j.compscitech.2005.06.007.
52. Mohan, T. P., and K. Kanny. Water barrier properties of nanoclay filled sisal fibre reinforced epoxy composites. *Composites Part A: Applied Science and Manufacturing* 42 (2011): 385–393.
53. Kim, H. J., and D. W. Seo. Effect of water absorption fatigue on mechanical properties of sisal textile-reinforced composites. *International Journal of Fatigue* 28 (2006): 1307–1314. doi:10.1016/j.ijfatigue.2006.02.018.
54. Dhakal, H. N., Z. Y. Zhang, and M. O. W. Richardson. Effect of water absorption on the mechanical properties of hemp fibre reinforced unsaturated polyester composites. *Composites Science and Technology* 67 (2007): 1674–1683. doi:10.1016/j.compscitech.2006.06.019.
55. Eskizeybek, V., H. Ulus, H. B. Kaybal, Ö. S. Şahin, and A. Avcı. Static and dynamic mechanical responses of $CaCO_3$ nanoparticle modified epoxy/ carbon fiber nanocomposites. *Composites Part B: Engineering* 140 (2018): 223–231. doi:10.1016/j.compositesb.2017.12.013.
56. Kumar, P. N., V. Nagappan, and C. Karthikeyan. Effects of fly ash, calcium carbonate fillers on mechanical, moisture absorption properties in poly vinyl chloride resin. *Materials Today: Proceedings* 16 (2019): 1219–1225. doi:10.1016/j. matpr.2019.05.217.
57. Li, L. H., Y. Chen, M.-Y. Lin, A. M. Glushenkov, B.-M. Cheng, and J. Yu. Single deep ultraviolet light emission from boron nitride nanotube film. *Applied Physics Letters* 97 (2010): 141104. doi:10.1063/1.3497261.
58. Chopra, N. G., and A. Zettl. Measurement of the elastic modulus of a multi-wall boron nitride nanotube. *Solid State Communications* 105 (1998): 297–300. doi:10.1016/S0038-1098(97)10125-9.
59. Kim, K., and J. Kim. Fabrication of thermally conductive composite with surface modified boron nitride by epoxy wetting method. *Ceramics International* 40 (2014): 5181–5189. doi:10.1016/j.ceramint.2013.10.076.

7

Hydrothermal Effect on Mechanical Properties of Nanofillers Embedded Natural Fiber Reinforced Polymer Composites

7.1 Introduction

Synthetic materials are not environmentally friendly materials, and they pollute the earth's environments. To reduce the environmental pollution, natural fiber polymer composites are used in place of the synthetic fiber reinforced polymer composite. The natural fibers are degradable and do not create pollution on the earth, they are also less expensive, have less cost and energy required for manufacturing, and have a better specific strength. The natural fibers, having better material properties than the synthetic fibers, are widely used in structural and semi-structural components. However, there are some drawbacks with natural fibers that include loss of strength at high temperatures and a hydrophilic nature. The polymer matrix has a better stability at high temperatures and a hydrophobic nature, which create a poor interface bond between the matrix and fibers. Hence, natural fiber reinforced polymer composites have impaired mechanical properties in hydrothermal environments. To improve the mechanical property and create strong bonds between the fibers and matrix, the nanoparticles or nanofillers impinge into the polymer matrix. The use of nanoparticles and nanofillers also improves the durability and reliability of the natural composites. In this chapter, we study how nanofillers or nanoparticles effect the mechanical properties of hydrothermally treated natural fiber reinforced polymer composites.

7.2 Natural Fiber Reinforced Composites

Natural fibers are obtained from plants in the form of cellulose fibrils, which are embedded in a lignin matrix. The biofiber structure is shown in Figure 7.1. Each biofiber has four layers of complex structures which

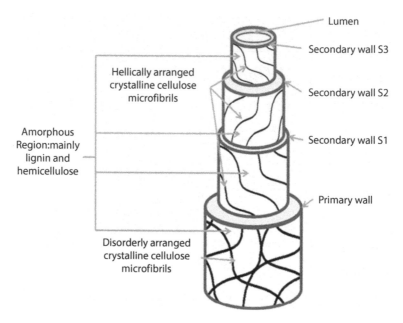

FIGURE 7.1
Structure of biofiber. (From Malvar, L.J., *Literature Review of Durability of Composites in Reinforced Concrete*, 1996.)

contain a primary and three secondary cell walls. The mechanical property of a biofiber depends on the thick middle layer of the secondary cell wall. It consists of a series of helically wound cellular microfibrils formed from long-chain cellulose molecules. Hemicelluloses, lignin, and cellulose are the main components of the cell wall. Lignin-hemicelluloses act as the matrix, while microfibrils (made up of cellulose molecules) act as fibers [1,2]. Other components include pectins, oil, and waxes [2,3]. Lumen is present in a natural fiber, making it a hollow structure, unlike synthetic fibers [4].

The biodegradation in composites occurs with the degradation of its individual constituents and loss of the interfacial strength between them. Figure 7.2 shows the cell wall polymers responsible for the properties of lingo cellulosics. Table 7.1 demonstrates the chemical composition, and Table 7.2 demonstrates the mechanical and physical properties of some natural fibers. Moreover, Table 7.2 compares the physical and mechanical properties between natural and synthetic fibers. It is observed that natural and synthetic fibers have elative mechanical and physical properties, so natural fibers are now the desired alternative of synthetic fibers in various applications. The use of natural fibers with polymer matrices (thermoset & thermoplastic) gives green composite materials with positive environmental profits with respect to the best utility of raw materials and disposability. Thus, some studies have been done in the last few years to replace the

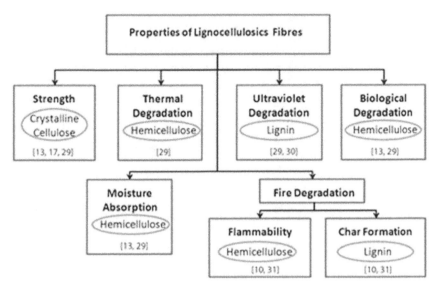

FIGURE 7.2
Cell wall polymers responsible for the properties of lignocellulosics. (From Azwa, Z.N. et al., *Mater. Des.*, 47, 424–442, 2013.)

TABLE 7.1

Chemical Composition of Selected Natural Fibers

Fiber Name	Cellulose (wt%)	Lignin (wt%)	Hemicellulose (wt%)	Pectin (wt%)
Coir	37	42		
Jute	61–71.5	12–13	17.9–22.4	0.2
Kenaf	54–57	21.5	8.13	0.6
Rice Husk	38–45		12–20	
Sisal	78	8	10	

Source: Ahmad, F. et al., *Macromol. Mater. Eng.*, 300, 10–24, 2015; Huda, M.S. et al., *Compos. Sci. Technol.*, 68, 424–432, 2008; Summerscales, J., et al., *Compos. Part Appl. Sci. Manuf.* 41, 1329–1335, 2010.

conventional synthetic composites with natural fiber composites. Fiber reinforced polymers are most likely jute, broom, instance, cotton, sisal, flax, and hemp with polyethylene [7], polyolefin [8], polystyrene [9], and epoxy [10]. The use of natural fibers is an alternative to traditional carbon and glass reinforcements for the improvement of a good thermal property, enhanced energy recovery, reduced tool wear, acoustic absorption, respiratory irritation, reduced dermal, and non-corrosive nature. Therefore, there is an increase in demand of natural fiber reinforced composites in various automobile applications [6], as mentioned in Table 7.3.

TABLE 7.2

Physical and Mechanical Properties of Selected Natural and Synthetic Fibers

	Fiber Name	Density (g cm³)	Diameter (mm)	Tensile Strength (MPa)	Specific Strength (S/ρ)	Tensile Modulus (GPa)	Specific Modulus (E/ρ)	Elongation at Break (%)
Natural fibers	Coir	1.2		175	146	4–6	3.3–5	30
	Jute	1.46	40–350	393–800	269–548	10–30	6.85–20.6	1.5–1.8
	Kenaf	1.45	70–250	930	641	53	36.55	1.6
	Sisal	1.45	50–300	530–640	366–441	9.4–22	6.5–15.2	3–7
Man-made fiber	Aramid	1.4	11.9	300	1916	124	86	2.5
	E-glass	2.55	<17	3400	1333	73	28	3.4
	HS Carbon	1.82	8.2	2550	1401	109	109	1.3
	S-glass	2.5		4580	1832	85	34	4.6

Source: Ahmad, F. et al., *Macromol. Mater. Eng.*, 300, 10–24, 2015; Rowell, R.M. et al., *Lignocellul-Plast Compos.*, 13, 23–51, 1997.

TABLE 7.3

Application of Natural Fiber Composites in Automobile

Manufacturer	Model	Applications
Audi	A2, A3, A4, A4-Avant, A6, A6-Avant, A8, Roadster, Coupe	Seat back, side and back door panel, boot lining, hat rack, and spare tire lining
BMW	3, 5, and 7 series	Door panels, headliner panel, boot lining, seat backs, noise insulation panels molded foot, and well linings
Ford	Mondeo CD 162, Focus	Door panels, B-pillar, and boot liner
Mercedes-Benz	Trucks	Internal engine cover, engine insulation, sun visor, interior, bumper, wheel box, and roof cover
Toyota	Brevis, Harrier, Celsior, RAUM	Door panels, seat backs, and spare tire cover
Volkswagen	Golf, Passat, Variant, Bora, Fox, Polo	Door panel, seat back, boot lid finish panel, and boot liner

Source: Ahmad, F. et al., *Macromol. Mater. Eng.*, 300, 10–24, 2015; Hill, K. et al., *Cent. Automot. Res.*, 112, 2012; Cristaldi, G. et al., *Woven Fabr. Eng.*, 17, 317–342, 2010; Koronis, G. et al., *Compo. Part B Eng.*, 44, 120–127, 2013.

7.3 Hydrothermal Effect

Fiber reinforced polymer composites have better corrosion resistance, high specific strength, and compatibility with other materials as compared to other metallic materials which attract the attention of designers selecting composite material for different engineering applications [17]. However, the performance and long-term durability of composite materials are of major concern and result in the hindrance of their wide applications. To improve the mechanical property of the composite materials, micro/nano inorganic filler materials are mixed with the epoxy matrix which helps to increase the application area of natural fiber composite materials. An inorganic nanofiller is easily manufactured and has a lesser cost, which increases the acceptability as potential fillers [18]. From the various research works, it is observed that micro-Al_2O_3 fillers improve the flexural strength impact and fracture toughness of carbon fiber reinforced polymer composites [19] and glass fiber reinforced polymer composites [20]. The thermo-mechanical properties of composite materials can be enhanced by incorporating CNT-Al_2O_3 hybrids in the epoxy matrix [21]. The nano-Al_2O_3 particles improve the bearing strength [22], storage modulus and glass transition temperature [23], flexural strength, thermal conductivity [24], Young's modulus [25], glass transition temperature, specific wear rate [26], and reduce the coefficient of thermal expansion [27]. Both nano- and micro-Al_2O_3 particles at different proportions improve the mechanical properties [28].

Researchers and engineers have added micro/nano-Al_2O_3 filler into the epoxy matrix, and they concluded that by the use of Al_2O_3 filler, the mechanical, impact, thermal conductivity, thermo-mechanical, glass transition temperature, wear, fracture toughness, and Young's modulus improve. Nayak et al. [29] have observed that the addition of nano-Al_2O_3 filler into the epoxy matrix increases the interlaminar shear strength. However, these materials are facing some difficulties in different environmental conditions like corrosive, water, high and low temperature, alkaline, and UV light exposure. In a hydrothermal or relative humidity/moisture environment, the composite materials absorb moisture/water molecules which are clustered in the voids present between the matrix and fibers or inside the matrix and make the bond with the hydroxyl group of the matrix. By this moisture absorption, the mechanical, physical, thermal, and electrical properties of the composite materials reduce [30,31]. The physical and chemical properties of the water-absorbed polymer changed. For example, the physical change of the epoxy polymer is basically plasticization and swelling. The chemical change of the epoxy is chain scission and hydrolysis [32–35]. The change in the physical property of the epoxy affected the glass transient temperature (T_g) of the glass fiber reinforced polymer composites. As the physical and chemical properties of the epoxy polymer change, its structures are also changed, which affect the mechanical property. The water-absorbed polymer changes the chemical, mechanical, and thermo-physical characteristics of the fiber reinforced polymer composites [36]. The presence of microcracks during matrix degradation of the mechanical properties under different environmental conditions can occur [37–39], therefore, it is challenging for the researchers and design engineers to retain the mechanical properties in a hydrothermal environment.

Adding of the nanofiller into the glass fiber reinforced polymer composites probably improves the interface strength [40]. The mechanical property is enhanced by the use of carbon base (multiwall carbon nano tube [MCNT], grapheme, carbon nanotube, and single wall carbon nano tube [SWCNT]) and inorganic (SiO_2, Al_2O_3, TiO_2, etc.) nanofillers. Okhawilai et al. [41] observed that the addition of nano-SiO_2 particles in epoxy-modified polybenzoxazine increases the modulus by 2.5 times. Inorganic fillers/particles are combined with the polymer matrix either in dispersed form, mechanically contacted, chemically bonded, or a combination of two or all [42]. Microcracks may deflect, pin, or blunt at the nanoparticle surface during their propagation, resulting in the improvement of fracture toughness [43]. Therefore, ductility and reliability of the composite materials at different environmental conditions are determined by the health of the interphase or interface.

The presence of a void in the polymer matrix composites plays an important role in the water absorption kinetics and mechanical properties of the glass fiber reinforced polymer. The resin burn off test (as per American Society for Testing and Materials [ASTM] D 3171-99 standard) is carried out to determine the volume fraction and fiber wt% of the void. According to the ASTM D 3171-99 standard, six numbers of specimens having a surface area of 25 mm × 25 mm each are taken into account for this analysis. A high accuracy weighing balance is used to measure weight, and a digital vernier caliper is used to measure the accurate dimensions of the samples. To burn, the epoxy of the composite material specimen is placed inside the muffle furnace at 575°C ± 10°C for 5 hours. Equation 7.1 shows the mathematical expression used to determine the void content [44,45],

$$V_v = 1 - \rho_c \left(\frac{w_f}{\rho_f} + \frac{w_m}{\rho_m} \right),$$ (7.1)

where V_v is the volume fraction of a void, ρ_c is the density of the composites, ρ_f is the density of the fiber, ρ_m is the density of the epoxy matrix, w_f is the weight fraction of the fiber, and w_m is the weight fraction of the epoxy matrix. Figure 7.3. shows the fiber wt% and void content of the control and nano-Al_2O_3-enhanced GFRP composites. During the compression molding, the epoxy drains out from the manufactured component/specimen, which slightly increased the fiber weight percentage after fabrication. Nayak et al. [44] observed that the glass fiber reinforced polymer with nanofiller has more void content than the control GFRP. This may be because gasses entrapped during the magnetic stirring and sonication were not removed properly during the compression molding.

Figure 7.4 shows the conditions of moisture in a polymer matrix. In a moisture/humid environment, the water penetrates into the natural fiber reinforced polymer composite and attaches to the hydrophilic group of the fiber, inhibiting intermolecular hydrogen bonding with the fiber and weakens the interfacial adhesion of the fiber-matrix. When swelling occurs, the degradation process starts, this creates stress at the interface regions, which develops a microcracking mechanism in the matrix around the swollen fibers, and

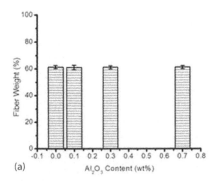
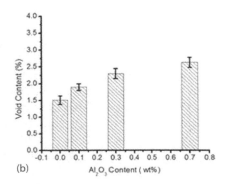

FIGURE 7.3
(a) Fiber weight% and (b) void content in the composites versus Al_2O_3 content (wt.%). (From Nayak, R.K. and Ray, B.C., *Polym. Bull.*, 74, 4175–4194, 2017.)

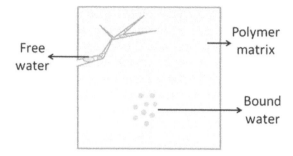

FIGURE 7.4
Free water and bound water in polymer matrix. (From Ahmad, F. et al., *Macromol. Mater. Eng.,* 300, 10–24, 2015.)

this promotes capillarity and transport via microcracking. The free water decreases and the bound water increases because water is highly absorbed. It is observed in a hemp fiber reinforced unsaturated polymer (hemp/UPE) composite, the water-soluble substances start to seep out from the fibers, and at the end leads to ultimate debonding between the matrix and fibers [47]. As the water-soluble substances seep out from the surface of the fibers, osmotic pressure pockets develop at the surface of the fibers, which leads to debonding between the fibers and matrix [48]. This process is summarized in Figure 7.7. After a long period, biological activities such as fungi growth degrade natural fibers [49]. The characteristics of the submerged natural composite fibers are influenced by the nature of the fiber material and matrix materials by relative humidity and manufacturing techniques, which determine factors such as porosity and fraction size of the fibers [47]. The water absorbed by the composite materials depends on various factors such as the orientation of reinforcement, area of the epoxy surface, surface property, the reaction between the water, the reaction between the matrix, temperature, permeability nature of the fiber, and fiber volume fraction [49].

At room temperature, the water absorption and desorption follow Fickian behavior, at the beginning, the water uptake process is linear, then slows and approaches stationary after a prolonged time. According to Fickian diffusion, the water is spreading from a higher concentration to a lower concentration caused by a concentration gradient [50]. The time of moisture saturation is reduced, and the uptake property increases at a higher submersion temperature. This non-Fickian behavior may be attributed to the difference in sorption behavior and the state of the water molecules existing in the composites [47,48]. The diffusion coefficient characterizes the ability of solvent molecules to move among the polymer segments. In a humid environment, the rise in temperature of the composite material starts the surface dissolving and peeling in large quantities with microcrack formations, which leads to the increase in the permeability coefficient [48]. Fickian and non-Fickian diffusion is shown in Figure 7.5a,b, respectively.

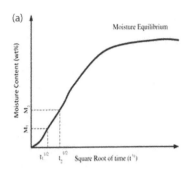 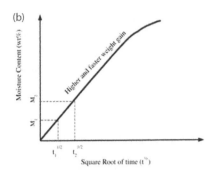

FIGURE 7.5
(a) Fickian diffusion at room temperature and (b) non-Fickian diffusion at elevated temperature. (From Dhakal, H.N. et al., *Compos. Sci. Technol.*, 67, 1674–1683, 2007.)

From the above study, it is observed that natural fiber/polymer composites lose their functionality in a humid environment. It is necessary to develop or manufacture a composite that can be used in a humid environment. It may be developed by proper selection of a natural fiber which does not react/absorb water molecules in a high humid environment.

The water absorbed in polymers consists of free water and bound water [49]. Figure 7.6 shows the free and bound water conditions in a polymer matrix. The bound water molecules are those water molecules that are

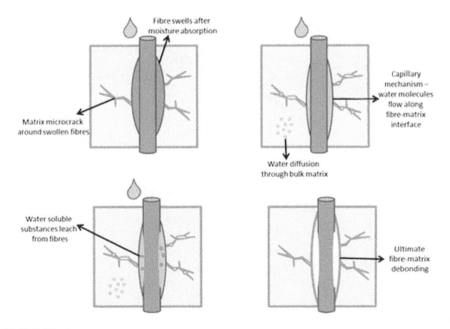

FIGURE 7.6
Effect of water on fiber-matrix interface. (From Azwa, Z.N. et al., *Mater. Des.*, 47, 424–42, 2013.)

bounded with a polar group of polymers, and the free water molecules are those water molecules that are able to move freely through voids.

As the natural fiber/polymer composites are present in the moisture environment, the water molecules start penetration into them and attach to a hydrophilic group of fibers, which leads to creating intermolecular hydrogen bonding with the fibers that reduce the interfacial adhesion between the matrix and fiber. When the swelling occurs, the degradation process starts, which creates stress at the interface regions and develops a microcracking mechanism in the matrix around the swollen fibers, and this promotes capillarity and transport via microcracking. The excess amounts of water absorbed by the composite lead to an increase in the inbound water and free water decreases. It is observed in a hemp fiber reinforced unsaturated polymer (hemp/UPE) composite that the water-soluble substances start to seep out from the fibers, and at the end leads to the ultimate debonding between the matrix and fibers [47]. As the water-soluble substances seep out from the surface of the fibers, osmotic pressure pockets develop at the surface of the fibers, which leads in debonding between the fibers and matrix [48]. This process is summarized in Figure 7.7. After a long period, biological activities such as fungi growth degrade natural fibers [49]. The characteristics of submerged natural composite fibers are influenced by the nature of the fiber material and matrix materials, by relative humidity and manufacturing techniques, which determine factors such as the porosity and the fraction size of the fibers [47]. The water absorbed by the composite materials depends on various factors

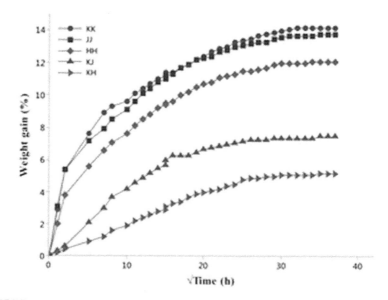

FIGURE 7.7
Water absorption behavior of different woven natural fiber composites. (From Maslinda, A.B. et al., *Compos. Struct.*, 167, 227–237, 2017.)

such as the orientation of reinforcement, area of epoxy surface, surface property, the reaction between water, the reaction between matrix, temperature, permeability nature of the fiber, and the fiber volume fraction [48]. From the above study, it is observed that natural fiber/polymer composites lose their functionality in a humid environment. It is necessary to develop or manufacture a composite that can be used in a humid environment. It is observed that when various woven composite materials are immersed into the tap water at room temperature, the moisture absorption percentage rate is a function of the square root of the time (in hours). Figure 7.7 shows the moisture absorption rate of various woven composites. It is observed from the water absorption curve, as the immersion time increases, the water absorption increases. This similar water absorption effect is studied in natural fiber reinforced polymer composites [51,52]. In the woven composite material, the water absorption is linear at the beginning and rapid with the penetration of water molecules, mainly in woven jute (JJ), kenaf (KK), and hemp (HH). The water absorption slowed down, and after a long time period, it reached the saturation. The moisture absorption is not following the Fickian behavior and seems more sigmoidal in nature. This may occur due to material lost, the water seeping out of the particles from the composites leads to weight loss. The kenaf fiber has a higher cellulose content, which may lead to an increase in the higher amount of water absorption than the other woven jute and hemp composite materials, as mentioned in Table 7.4

Cellulose is a hygroscopic and main constituent of plants fibers, due to the cellulose hygroscopic property, it absorbs moisture in the largest amount [51]. The role of the resin matrix in the composite material is to protract the fibers and help to transfer the load from one fiber to another [54]. When the composite is immersed into the moisture environment, fibers start swelling, which produces the microcracking in a brittle thermosetting resin. As kenaf fibers contain high cellulose, which helps to absorb more amounts of water, this leads to the development of stress in the fiber matrix interphase that fails the composite materials. As microcracks increase, the water transport via capillary action in the composite material is also increased [47]. The water molecules continuously attack the interface, while moving with capillarity, cracks lead to debonding between the fiber and matrix.

TABLE 7.4

Chemical Composition of the Natural Fiber Reinforcement

Constituent/Fiber	Cellulose (wt%)	Hemicellulose (wt%)	Lignin (wt%)	Waxes (wt%)
Hemp	68	15	10	0.8
Jute	61–71	14–20	12–13	0.5
Kenaf	72	20.3	9	–

Source: Faruk, O. et al., *Prog. Polym. Sci.*, 37, 1552–1596, 2012. doi:10.1016/j.progpolymsci.2012.04.003.

7.3.1 Mechanical Properties

7.3.1.1 Tensile Strength

The tensile test under dry and wet conditions is carried out on woven and interwoven hybrid composites to determine their elastic modulus, strength, and strain to failure. At different immersion times for the wet specimen, a tensile test is carried out to obtain the degradation response. Table 7.5 shows the tensile properties. From Figure 7.8, it is observed that when the tensile test is carried out on a hybrid interwoven composite in dry and saturated conditions, failure strain, elastic modulus, and tensile strength are increased. In a dry condition, a hybrid kenaf/jute interwoven composite (KJ) shows the

TABLE 7.5

Tensile Properties of Dry and Saturated Woven Fiber Reinforced Polymer Composites

		Specimens				
Properties		**HH**	**JJ**	**KK**	**HJ**	**KH**
Tensile strength (MPa)	Dry	76 ± 5	77 ± 3	80 ± 4	89 ± 3	38 ± 1
	Saturated	25 ± 3	24 ± 1	20 ± 4	25 ± 2	26 ± 1
Tensile modulus (GPa)	Dry	5.3 ± 0.5	5.7 ± 0.3	6 ± 0.5	7.4 ± 0.5	7.3 ± 0.3
	Saturated	1.4 ± 0.1	1.3 ± 0.1	1 ± 0.2	1.4 ± 0.2	1.5 ± 0.1
Tensile strain (%)	Dry	2.9 ± 0.6	2.4 ± 0.2	2.2 ± 0.3	3.1 ± 0.5	3.5 ± 0.3
	Saturated	5.5 ± 0.1	5.3 ± 0.3	5.1 ± 0.2	5.8 ± 0.6	6.0 ± 0.2

Source: Maslinda, A.B. et al., *Compos. Struct.*, 167, 227–237, 2017.

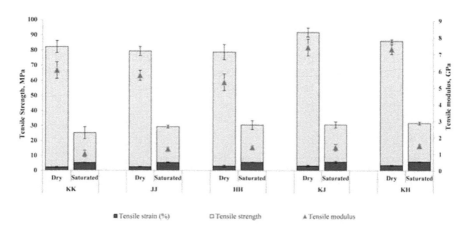

FIGURE 7.8

Tensile properties of various woven composites in dry and saturated conditions. (From Maslinda, A.B. et al., *Compos. Struct.*, 167, 227–237, 2017.)

failure strain is increased by 41%, elastic modulus is increased by 23%, and tensile strength is increased by 11% in comparison with the KK composite; and the failure strain is increased by 29%, elastic modulus is increased by 30%, and tensile strength is increased by 16% in comparison with the JJ composite. In a hybrid kenaf/hemp interwoven composite (KH), failure strain, elastic modulus, and tensile strength are increased by 59%, 22%, and 4%, respectively, over a KK composite and increased by 21%, 38%, and 9%, respectively, in comparison with a hemp woven composite. The hybrid interwoven composite in the saturated condition follows a similar trend where the failure strain, elastic modulus, and tensile strength increased. In a saturated condition, a hybrid KJ composite shows the failure strain is increased by 14%, elastic modulus is increased by 40%, and tensile strength is increased by 25% in comparison with the KK composite, and the failure strain is increased by 9%, elastic modulus is increased by 8%, and tensile strength is increased by 4% in comparison with the JJ composite. In a hybrid KH composite, the failure strain, elastic modulus, and tensile strength are increased by 18%, 50%, and 30%, respectively, in comparison with a KK composite and increased by 9%, 7%, and 4%, respectively, in comparison with a hemp composite. From the above observations, it is observed that kenaf/jute and kenaf/hemp hybrid interwoven composites have the ability to resist deformation and breaking under tensile loading in comparison with individual kenaf, jute, and hump woven composite. The property of hybrid composites depends on the reinforcement selection so the desired property is achieved by proper selection of the reinforcement fibers [54]. The kenaf woven fiber is used to hybridize with jute and hemp woven fibers to achieve the outstanding mechanical property of hybrid composites.

7.3.1.2 Flexural Strength

To replace the metallic materials from a hydrothermal environment, it is necessary to obtain the reliability and durability of nano-composite material in a hydrothermal environment. Matrix toughness and interface strength are very important for nano-composites in a hydrothermal environment. The interface strength can be tailored through the flexural test, and it is evaluated as per ASTM D7264 standard.

The flexural strength (σ_F), modulus (E_F), and strain to failure (ε_F) are calculated as per the equation given below [56]:

$$\sigma_F = \frac{3P_{max}L}{2wt^2} \quad \text{when} \quad \frac{L}{t} \leq 16 \tag{7.2}$$

$$E_F = \frac{mL^3}{4wt^3} \tag{7.3}$$

$$\varepsilon_F = \frac{6dt}{L^2},$$ (7.4)

where L is the span length of the sample, w is the width, and t is the thickness of the sample, P_{max} is the maximum load applied before failure, m is the slope of the initial portion of the load versus the displacement curve, and d is the maximum bending before failure.

Table 7.6 summarizes the results obtained from the flexural test in dry- and water-aged samples at different durations. As a result of hybridization, excellent flexural properties were achieved. In dry and saturated conditions, flexural properties of kenaf/jute and kenaf/hemp hybrid interwoven composites as shown in Figure 7.9 are higher than the individual kenaf, jute, and hemp interweaved fiber composites. It is observed that when the flexural test is carried out on hybrid interwoven composites in the dry and saturated conditions, the failure strain, flexural modulus, and flexural strength are increased. In a dry condition, the hybrid KJ composite shows the failure strain is increased by 16%, flexural modulus is increased by 12%, and flexural strength is increased by 22% in comparison with the KK composite; and the failure strain is increased by 16%, flexural modulus is increased by 26%, and flexural strength is increased by 39% in comparison with the JJ composite. In the hybrid KH composite, the failure strain, flexural modulus, and flexural strength are increased by 20%, 4%, and 17%, respectively, over the KK composite; and increased by 5%, 23%, and 33%, respectively, in comparison with the hemp woven composite. The hybrid interwoven composite in the saturated condition follows a similar trend, where the failure strain, flexural modulus, and flexural strength increased. In a saturated condition, the hybrid KJ composite shows the failure strain is increased by 13%, flexural modulus is increased by 27%, and flexural strength is increased by 41% in comparison with the KK composite, and the failure strain is increased by 10%, flexural modulus is increased by 3%, and flexural strength is increased by 18% in comparison with the JJ composite. In the hybrid KH composite,

TABLE 7.6

Flexural Strength of Dry and Saturated Composites

Specimens	Dry (MPa)	Wet (MPa)			
	0 h	2 h	5 h	15 h	24 h
JJ	68.5 ± 5.3	54 ± 3.6	50.2 ± 4.7	44 ± 5.9	38.3 ± 2.6
KK	77.6 ± 2.9	52.1 ± 2.5	49.9 ± 1.3	42.9 ± 1.3	35.8 ± 1.2
KJ	95 ± 3.4	62.6 ± 1.7	51.5 ± 1.6	46.3 ± 2.2	39 ± 2.7
	0 h	2 h	10 h	24 h	48 h
HH	68.2 ± 4.9	55.5 ± 1.2	48.3 ± 1.4	39 ± 2.4	29.8 ± 1.4
KH	90.8 ± 6.3	63.3 ± 5.5	48.4 ± 1.1	42.6 ± 4.4	30.6 ± 4.3

Source: Maslinda, A.B. et al., *Compos. Struct.*, 167, 227–237, 2017.

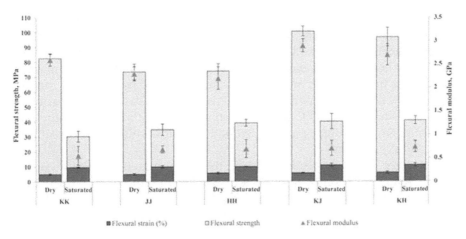

FIGURE 7.9
Flexural properties of various woven composites in dry and saturated conditions. (From Maslinda, A.B. et al., *Compos. Struct.*, 167, 227–237, 2017.)

the failure strain, flexural modulus, and flexural strength are increased by 13%, 32%, and 44%, respectively, in comparison with the KK composite; and increased by 10%, 6%, and 2%, respectively, in comparison with the hemp woven composite. From the above observations, it is observed that kenaf/jute and kenaf/hemp hybrid interwoven composites are stronger and more rigid in comparison with the individual kenaf, jute, and hump woven composites. This indicates that hybridization with high strength fibers, such as kenaf, yields a material with a better flexural performance.

7.3.2 Thermal Properties

The thermal property of a glass fiber reinforced polymer composite is very critical in terms of reliability and sustainability. Figure 7.10 shows the temperature modulated differential scanning calorimeter (TMDSC-822, Mettler Toledo 821) to evaluate the glass transition temperature of the control and

FIGURE 7.10
FTMDSC equipment (Mettler Toledo) used for glass transition temperature.

nano-GFRP composites. According to the TMDSC principle, the heat required to raise the temperature in both the reference sample and the working sample is a function of time and temperature. During the experiment, the temperature of both the reference sample and working sample are maintained equally. The time constant and heat flow are calibrated by the use of indium (In) and zinc (Zn). The experiment is carried out in a nitrogen inert gas environment from 25°C to 150°C with a rise in temperature with the heating rate 10°C/min. An aluminum pan is used and heat flow calibration is required before the experiment.

The weight loss is determined by the thermogravimetric analysis curves, and the material decomposition at a certain temperature is identified by a derivative thermogravimetric curve [57]. Figure 7.11 shows the typical thermogravimetric graph of natural fiber for the decomposition process. The thermal decomposition processes of different lignocellulosic fibers have very similar thermogravimetric analysis and derivative thermogravimetric curves due to their similar characteristics. Around 60% of natural fibers are thermally decomposed between a 215°C–310°C thermal range of 160–170 kJ/mol activation energy [60]. The decomposition temperature is used to evaluate the thermal degradation behavior. For studied fibers in [61], the decomposition temperature is at 290°C–490°C. Based on [62], when the degradation experiment is carried out on natural fibers, at around a 260°C temperature, an emission of volatile components initiates with dehydration and oxidation decomposition occurs, which creates rapid weight loss which leads to char as the temperature rises. Early decomposition shows less thermal stability [61].

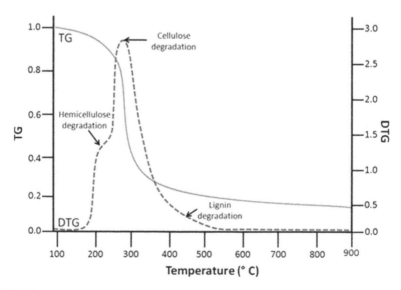

FIGURE 7.11
Typical thermogravimetric decomposition process of natural fibers. (From Suardana, N.P.G. et al., *Mater. Des.*, 32, 1990–1999, 2011.)

TABLE 7.7

The Main Stages of Weight Loss of Natural Fibers

Stage 1	Stage 2	Stage 3
50°C–100°C: Evaporation of moisture in the fibers	200°C–300°C: Decomposition of hemicelluloses	400°C–500°C: Weight loss due to lignin and cellulose degradation
300°C: Corresponds to the thermal decomposition of hemicellulose and the glycosidic link of cellulose	360°C: Corresponds to the thermal decomposition of α-cellulose	200°C–500°C, max at 350°C: Lignin peak is wider and appears superposed on the other two peaks
250°C–300°C: Characteristic of low molecular weight components, such as hemicelluloses	300°C–400°C: Corresponds to the thermal degradation of cellulose	Near 420°C: Due to lignin decomposition
220°C–315°C: Pyrolysis of hemicelluloses	315°C–400°C: Pyrolysis of cellulose	160°C–900°C: Pyrolysis of lignin
97°C: Attributed to water loss	325°C: Attributed to protein degradation	Major thermal decomposition of the composites began at about 260°C and beyond 390°C the rate of decomposition was slow

Table 7.7 shows the three main stages of weight loss of natural fibers during degradation experiments by some researchers.

7.3.3 Wear Resistance

The natural fiber reinforced polymer composite is eco-friendly in nature, easily available, and costs less, as these synthetic fibers are replaced by natural fibers in reinforced composite materials [63]. Therefore, for the last few years, the automobile, packaging, and construction industries have shown great interest in developing new natural FRPC materials. The natural fibers are eco-friendly, cheap, easily manufactured, and available, which attract researchers to use them as reinforcement in composite materials to determine their feasible application and extend the requirements in a tribological application.

In this study, natural fiber reinforced polymer composites are reported in reference to their test parameters, matrix polymers, and types and treatment of fibers. The natural fiber and conventional fiber composite properties are compared. From the above studies, it is observed that the natural fiber composites can improve tribological properties. The tribological properties of composites depend on the fiber orientation and treatment, where normally oriented and treated fibers show better wear and friction behavior. This review attempts to assess the effect of test parameters, including normal load and sliding speed, on tribal properties, and the results vary depending on the type of reinforcement. Generally, natural fiber reinforced composite materials have positive economic and eco-friendly aspects in nature, as well as their good tribological properties are used in various applications areas.

Tribology mainly has two important phenomena, friction and wear, which arise during relative motion of the solid surface, and they try to wear materials and dissipate energy [55–59]. In general, there are various ways to improve the tribological behavior of neat polymer alloys. The popular method is embedding fibers in polymers to make composites [60–66]. Carbon fiber, glass fiber, and aramid fibers are well known fibers that are commonly used by composite industries. Weight loss and volume loss are the two methods to determine the wear rate. A tribotest is carried out to obtain the weight loss by calculating the initial weight and final weight difference. The following equation shows the correlation between mass loss and volume loss:

$$\text{Volume mass (mm}^3/\text{N} \cdot \text{m)} = [\text{mass loss (g)}/\text{density (g/cm}^3)]/[\text{load (N)}$$

$$\times \text{sliding distance (m)}] \times 1000.$$

Figure 7.12 illustrates the coefficient of the friction of jute fabric and neat reinforced polypropylene (PP) composites effected by the different applied loads (10, 20, and 30 N) and sliding speeds (1, 2, and 3 m/s). Figure 7.13 illustrates the variation of the specific wear rate of the jute fabric and neat reinforced PP composites effected by different applied loads (10, 20, and 30 N) and sliding speeds (1, 2, and 3 m/s). An on-disc tester is used to determine the wear against a hardened steel disc (62 HRC) with 1.6 μm Ra surface roughness. The tribological property of the composite material is affected by temperature, as the temperature rises, the surface becomes soft, which can increase surface wear.

At a high applied load, the coefficient of friction is reduced, which indicates that the temperature increased between the counter face and composites. This may be the likely mechanism behind the reduction in COF values at a higher load. Figure 7.13 illustrates the variation of the specific wear rate of

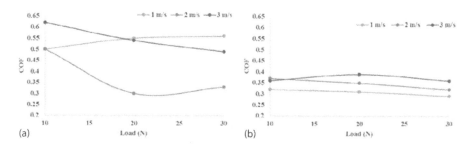

FIGURE 7.12
Friction coefficient against applied load for (a) neat PP and (b) jute/PP. (From Yallew, T.B. et al., *Procedia. Eng.*, 97, 402–411, 2014.)

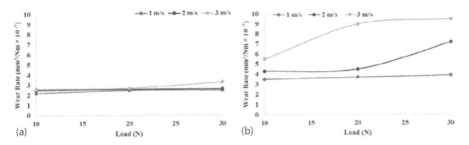

FIGURE 7.13
Variation of specific wear rate against applied load for (a) neat PP and (b) jute/PP. (From Yallew, T.B. et al., *Procedia. Eng.*, 97, 402–411, 2014.)

woven jute fabric and neat reinforced polypropylene composites effected by different applied loads and sliding speeds. At a high applied load, the specific wear rate is increased for both the jute and neat polypropylene fabrication composites [65]. At a high sliding speed, neat and jute polypropylene composites show poor wear performance. In some research, nano- and micron-size particles are used in hybrid polymer composites to improve the tribological and mechanical properties of the natural fiber composites because they do not show satisfactory mechanical properties. The SiC particles are added into the hybrid jute epoxy composites, which lead to an increase in mechanical behavior, as shown in Table 7.8, which also helps to increase the tribological properties [64].

Figure 7.14 shows improvement in the wear resistance property in the jute/epoxy composites by adding alumina and SiC [67]. As per ASTM G99 standard pin-on-disc wear test is carried out with a sliding distance of 1800 m at a constant speed to determine the dry sliding wear. The maximum weight loss of 21.8×10^{-3} g for the jute/epoxy composite is obtained at 3 m/s sliding speed and 50 N normal force, while the addition of Sic and Al_2O_3 in the jute reinforced polymer composite gives the drastic reduction in wear loss that is obtained, where the weight loss is 2.9×10^{-3} and 3.3×10^{-3} g in the presence of 15 wt% Al_2O_3 and SiC, respectively.

TABLE 7.8

Mechanical Properties of the Composites

Composite	Hardness (Hv)	Tensile Strength (MPa)	Flexural Strength (MPa)
Epoxy/20 wt % jute	57	302	312
Epoxy/30 wt% jute	59	331	345
Epoxy/40 wt% jute	63	349	368
Epoxy/40 wt% jute/10 wt% Sic	83	304	357
Epoxy/40 wt% jute/20 wt% Sic	86	279	353

Source: Behera, F.K. and Murali, M.G.B. *Int. J. Innov. Res. Sci. Eng. Technol.*, 3, 10–18, 2015.

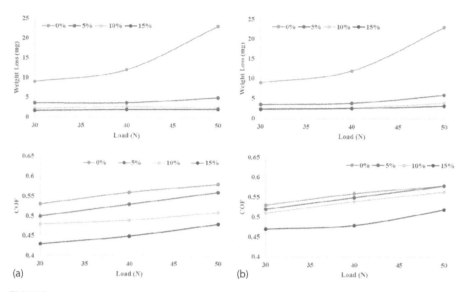

FIGURE 7.14
Wear loss and COF versus normal load for constant velocity of 3 m/s: (a) for Al_2O_3 and (b) for SiC. (From Ahmed et al., *Mater. Design*, 36, 306–315, 2012.)

References

1. Dittenber, D. B., and H. V. GangaRao. Critical review of recent publications on use of natural composites in infrastructure. *Composites Part A: Applied Science and Manufacturing* 43 (2012): 1419–1429.
2. John, M. J., and S. Thomas. Biofibers and biocomposites. *Carbohydrate Polymers* 71 (2008): 343–364.
3. Wong, K. J., B. F. Yousif, and K. O. Low. The effects of alkali treatment on the interfacial adhesion of bamboo fibres. *The Proceedings of the Institution of Mechanical Engineers, Part L: Journal of Materials: Design and Applications* 224 (2010): 139–148.
4. Liu, K., H. Takagi, R. Osugi, and Z. Yang. Effect of lumen size on the effective transverse thermal conductivity of unidirectional natural fiber composites. *Composites Science and Technology* 72 (2012): 633–639.
5. Azwa, Z. N., B. F. Yousif, A. C. Manalo, and W. Karunasena. A review on the degradability of polymeric composites based on natural fibres. *Materials & Design* 47 (2013): 424–442. doi:10.1016/j.matdes.2012.11.025.
6. Ahmad, F., H. S. Choi, and M. K. Park. A review: Natural fiber composites selection in view of mechanical, light weight, and economic properties. *Macromolecular Materials and Engineering* 300 (2015): 10–24.
7. Roumeli, E., Z. Terzopoulou, E. Pavlidou, K. Chrissafis, E. Papadopoulou, E. Athanasiadou, et al. Effect of maleic anhydride on the mechanical and thermal properties of hemp/high-density polyethylene green composites. *Journal of Thermal Analysis and Calorimetry* 121 (2015): 93–105.

8. Russo, P., G. Simeoli, D. Acierno, and V. Lopresto. Mechanical properties of virgin and recycled polyolefin-based composite laminates reinforced with jute fabric. *Polymer Composites* 36 (2015): 2022–2029.

9. Oumer, A.N., and D. Bachtiar. Modeling and experimental validation of tensile properties of sugar palm fiber reinforced high impact polystyrene composites. *Fibers and Polymers* 15 (2014): 334–339.

10. Irawan, A. P., and I. W. Sukania. Tensile strength of banana fiber reinforced epoxy composites materials. *Applied Mechanics and Materials* 776 (2015): 260–263.

11. Hill K., B. Swiecki, and J. Cregger. The bio-based materials automotive value chain. *Center for Automotive Research* (2012): 112.

12. Cristaldi, G., A. Latteri, G. Recca, and G. Cicala. Composites based on natural fibre fabrics. *Woven Fabric Engineering* 17 (2010): 317–342.

13. Koronis, G., A. Silva, and M. Fontul. Green composites: A review of adequate materials for automotive applications. *Composites Part B: Engineering* 44 (2013): 120–127.

14. Huda, M. S., L. T. Drzal, A. K. Mohanty, and M. Misra. Effect of fiber surface-treatments on the properties of laminated biocomposites from poly (lactic acid) (PLA) and kenaf fibers. *Composites Science and Technology* 68 (2008): 424–432.

15. Summerscales, J., N. P. Dissanayake, A. S. Virk, and W. Hall. A review of bast fibres and their composites. Part 1–Fibres as reinforcements. *Composites Part A: Applied Science and Manufacturing* 41 (2010): 1329–1335.

16. Rowell, R. M., A. R. Sanadi, D. F. Caulfield, and R. E. Jacobson. Utilization of natural fibers in plastic composites: Problems and opportunities. *Lignocellul-Plast Composites* 13 (1997): 23–51.

17. Malvar, L. J. Literature review of durability of composites in reinforced concrete, Special Publications SP-2008-SHR, Naval Facilities Engineering Service Center, Port Hueneme, California (1996): 26. 1996.

18. Alexandre, M., and P. Dubois. Polymer-layered silicate nanocomposites: Preparation, properties and uses of a new class of materials. *Materials Science and Engineering R: Reports* 28 (2000): 1–63. doi:10.1016/S0927-796X(00)00012-7.

19. Wang, Z., X. Huang, L. Bai, R. Du., Y. Liu, Y. Zhang. et al. Effect of micro-Al_2O_3 contents on mechanical property of carbon fiber reinforced epoxy matrix composites. *Composites Part B: Engineering* 91 (2016): 392–398. doi:10.1016/j.compositesb.2016.01.052.

20. Nayak, R. K., A. Dash, and B. C. Ray. Effect of epoxy modifiers ($Al_2O_3/SiO_2/TiO_2$) on mechanical performance of epoxy/glass fiber hybrid composites. *Procedia Materials Science* 6 (2014): 1359–1364. doi:10.1016/j.mspro.2014.07.115.

21. Li, W., A. Dichiara, J. Zha, Z. Su., and J. Bai. On improvement of mechanical and thermo-mechanical properties of glass fabric/epoxy composites by incorporating CNT–Al_2O_3 hybrids. *Composites Science and Technology* 103 (2014): 36–43. doi:10.1016/j.compscitech.2014.08.016.

22. Asi, O. An experimental study on the bearing strength behavior of Al_2O_3 particle filled glass fiber reinforced epoxy composites pinned joints. *Composite Structures* 92 (2010): 354–363. doi:10.1016/j.compstruct.2009.08.014.

23. Omrani, A., and A. A. Rostami. Understanding the effect of nano-Al_2O_3 addition upon the properties of epoxy-based hybrid composites. *Materials Science and Engineering: A* 517 (2009): 185–190. doi:10.1016/j.msea.2009.03.076.

24. Hu, Y., G. Du., and N. Chen. A novel approach for Al_2O_3/epoxy composites with high strength and thermal conductivity. *Composites Science and Technology* 124 (2016): 36–43. doi:10.1016/j.compscitech.2016.01.010.

25. Moreira, D. C., L. A. Sphaier, J. M. L. Reis, and L. C. S. Nunes. Determination of Young's modulus in polyester-Al_2O_3 and epoxy-Al_2O_3 nanocomposites using the digital image correlation method. *Composites Part A: Applied Science and Manufacturing* 43 (2012): 304–309. doi:10.1016/j.compositesa.2011.11.005.
26. Shi, G., M. Q. Zhang, M. Z. Rong, B. Wetzel, and K. Friedrich. Sliding wear behavior of epoxy containing nano-Al_2O_3 particles with different pretreatments. *Wear* 256 (2004): 1072–1081. doi:10.1016/S0043-1648(03)00533-7.
27. Jiang, W., F. L. Jin, and S.-J. Park. Thermo-mechanical behaviors of epoxy resins reinforced with nano-Al_2O_3 particles. *Journal of Industrial and Engineering Chemistry* 18 (2012): 594–596. doi:10.1016/j.jiec.2011.11.140.
28. Hussain, M., A. Nakahira, and K. Niihara. Mechanical property improvement of carbon fiber reinforced epoxy composites by Al_2O_3 filler dispersion. *Materials Letters* 26 (1996): 185–191. doi:10.1016/0167-577X(95)00224-3.
29. Nayak, R. K., D. Rathore, B. C. Routara, and B. C. Ray. Effect of nano Al_2O_3 fillers and cross head velocity on interlaminar shear strength of glass fiber reinforced polymer composite. *International Journal of Plastics Technology* (2016): 1–11. doi:10.1007/s12588-016-9158-z.
30. Gonon, P., A. Sylvestre, J. Teysseyre, and C. Prior. Combined effects of humidity and thermal stress on the dielectric properties of epoxy-silica composites. *Materials Science and Engineering: B* 83 (2001): 158–164. doi:10.1016/S0921-5107(01)00521-9.
31. Maggana C., and P. Pissis. Water sorption and diffusion studies in an epoxy resin system. *Journal of Polymer Science Part B: Polymer Physics* 37 (1999): 1165–1182. doi:10.1002/(SICI)1099-0488(19990601)37:11<1165::AID-POLB11>3.0.CO;2-E.
32. Verghese, K. N. E., M. D. Hayes, K. Garcia, C. Carrier, J. Wood, J. R. Riffle, et al. Influence of matrix chemistry on the short term, hydrothermal aging of vinyl ester matrix and composites under both isothermal and thermal spiking conditions. *Journal of Composite Materials* 33 (1999): 1918–1938. doi:10.1177/002199839903302004.
33. De'Nève B., and M. E. R. Shanahan. Water absorption by an epoxy resin and its effect on the mechanical properties and infra-red spectra. *Polymer* 34 (1993): 5099–5105. doi:10.1016/0032-3861(93)90254-8.
34. Xiao, G. Z., M. Delamar, and M. E. R. Shanahan. Irreversible interactions between water and DGEBA/DDA epoxy resin during hygrothermal aging. *Journal of Applied Polymer Science* 65 (1997): 449–458. doi:10.1002/(SICI)1097-4628(19970718)65:3<449::AID-APP4>3.0.CO;2-H.
35. Ray, B. C. Temperature effect during humid ageing on interfaces of glass and carbon fibers reinforced epoxy composites. *Journal of Colloid and Interface Science* 298 (2006): 111–117. doi:10.1016/j.jcis.2005.12.023.
36. Yilmaz, T., and T. Sinmazcelik. Effects of hydrothermal aging on glass–fiber/polyetherimide (PEI) composites. *Journal of Materials Science* 45 (2009): 399–404. doi:10.1007/s10853-009-3954-1.
37. Gautier, L., B. Mortaigne, and V. Bellenger. Interface damage study of hydrothermally aged glass-fibre-reinforced polyester composites. *Composites Science and Technology* 59 (1999): 2329–2337. doi:10.1016/S0266-3538(99)00085-8.

38. Hodzic, A., J. K. Kim, A. E. Lowe, and Z. H. Stachurski. The effects of water aging on the interphase region and interlaminar fracture toughness in polymer–glass composites. *Composites Science and Technology* 64 (2004): 2185–2195. doi:10.1016/j.compscitech.2004.03.011.
39. Ellyin, F., and R. Maser. Environmental effects on the mechanical properties of glass-fiber epoxy composite tubular specimens. *Composites Science and Technology* 64 (2004): 1863–1874. doi:10.1016/j.compscitech.2004.01.017.
40. Fan, X. J., S. W. R. Lee, and Q. Han. Experimental investigations and model study of moisture behaviors in polymeric materials. *Microelectronics Reliability* 49 (2009): 861–871. doi:10.1016/j.microrel.2009.03.006.
41. Okhawilai, M., I. Dueramae, C. Jubsilp, and S. Rimdusit. Effects of high nano-SiO_2 contents on properties of epoxy-modified polybenzoxazine. *Polymer Composites* 2015: n/a–n/a. doi:10.1002/pc.23807.
42. Kinloch, A. J., K. Masania, A. C. Taylor, S. Sprenger, and D. Egan. The fracture of glass-fibre-reinforced epoxy composites using nanoparticle-modified matrices. *Journal of Materials Science* 43 (2007): 1151–1154. doi:10.1007/s10853-007-2390-3.
43. Park, H., L. Su-Jin, K. Yoon-Jeong, C.-I. Jang, and J.-P. Won. Mechanical properties and microstructures of GFRP rebar after long-term exposure to chemical environments. *Polymer Composites* 2007;15: 403–408.
44. Nayak, R. K., K. K. Mahato, and B. C. Ray. Water absorption behavior, mechanical and thermal properties of nano TiO_2 enhanced glass fiber reinforced polymer composites. *Composites Part A: Applied Science and Manufacturing* 90 (2016): 736–747. doi:10.1016/j.compositesa.2016.09.003.
45. Nayak, R. K., K. K. Mahato, B. C. Routara, and B. C. Ray. Evaluation of mechanical properties of Al_2O_3 and TiO_2 nano filled enhanced glass fiber reinforced polymer composites. *Journal of Applied Polymer Science* 133 (2016) doi:10.1002/app.44274.
46. Nayak, R. K., and B. C. Ray. Water absorption, residual mechanical and thermal properties of hydrothermally conditioned nano-Al_2O_3 enhanced glass fiber reinforced polymer composites. *Polymer Bulletin* 74 (2017): 4175–4194.
47. Dhakal, H. N., Z. Y. Zhang, and M. O. W. Richardson. Effect of water absorption on the mechanical properties of hemp fibre reinforced unsaturated polyester composites. *Composites Science and Technology* 67 (2007): 1674–1683. doi:10.1016/j.compscitech.2006.06.019.
48. Joseph, P. V., M. S. Rabello, L. H. C. Mattoso, K. Joseph, and S. Thomas. Environmental effects on the degradation behaviour of sisal fibre reinforced polypropylene composites. *Composites Science and Technology* 62 (2002): 1357–1372. doi:10.1016/S0266-3538(02)00080-5.
49. Chen, H., M. Miao, and X. Ding. Influence of moisture absorption on the interfacial strength of bamboo/vinyl ester composites. *Composites Part A: Applied Science and Manufacturing* 40 (2009): 2013–2019. doi:10.1016/j.compositesa.2009.09.003.
50. Bao, L.-R., A. F. Yee, and C. Y. C. Lee. Moisture absorption and hygrothermal aging in a bismaleimide resin. *Polymer* 42 (2001): 7327–7333. doi:10.1016/S0032-3861(01)00238-5.

51. Akil, H. M., M. F. Omar, A. A. M. Mazuki, S. Safiee, Z. A. M. Ishak, and A. Abu Bakar. Kenaf fiber reinforced composites: A review. *Materials & Design* 32 (2011): 4107–4121. doi:10.1016/j.matdes.2011.04.008.
52. Salleh, Z., Y. M. Taib, K. M. Hyie, M. Mihat, M. N. Berhan, and M. A. A. Ghani. Fracture toughness investigation on long kenaf/woven glass hybrid composite due to water absorption effect. *Procedia Engineering* 41 (2012): 1667–1673. doi:10.1016/j.proeng.2012.07.366.
53. Faruk, O., A. K. Bledzki, H.-P. Fink, and M. Sain. Biocomposites reinforced with natural fibers: 2000–2010. *Progress in Polymer Science* 37 (2012): 1552–1596. doi:10.1016/j.progpolymsci.2012.04.003.
54. Dan-Mallam Y., M. Z. Abdullah, and P. S. M. M. Yusoff. The effect of hybridization on mechanical properties of woven kenaf fiber reinforced polyoxymethylene composite. *Polymer Composites* 35 (2014): 1900–1910. doi:10.1002/pc.22846.
55. Maslinda, A. B., M. S. Abdul Majid, M. J. M. Ridzuan, M. Afendi, and A. G. Gibson. Effect of water absorption on the mechanical properties of hybrid interwoven cellulosic-cellulosic fibre reinforced epoxy composites. *Composite Structures* 167 (2017): 227–237. doi:10.1016/j.compstruct.2017.02.023.
56. Dong, C., and I. J. Davies. Optimal design for the flexural behaviour of glass and carbon fibre reinforced polymer hybrid composites. *Materials & Design* 37 (2012): 450–457.
57. Suardana, N. P. G., M. S. Ku., and J. K. Lim. Effects of diammonium phosphate on the flammability and mechanical properties of bio-composites. *Materials & Design* 32 (2011): 1990–1999. doi:10.1016/j.matdes.2010.11.069.
58. Lee, S.-H., and S. Wang. Biodegradable polymers/bamboo fiber biocomposite with bio-based coupling agent. *Composites Part A: Applied Science and Manufacturing* 37 (2006): 80–91. doi:10.1016/j.compositesa.2005.04.015.
59. di Franco, C. R., V. P. Cyras, J. P. Busalmen, R. A. Ruseckaite, and A. Vázquez. Degradation of polycaprolactone/starch blends and composites with sisal fibre. *Polymer Degradation and Stability* 86 (2004): 95–103. doi:10.1016/j.polymdegradstab.2004.02.009.
60. Yao, F., Q. Wu., Y. Lei, W. Guo, and Y. Xu. Thermal decomposition kinetics of natural fibers: Activation energy with dynamic thermogravimetric analysis. *Polymer Degradation and Stability* 93 (2008): 90–98. doi:10.1016/j.polymdegradstab.2007.10.012.
61. Methacanon, P., U. Weerawatsophon, N. Sumransin, C. Prahsarn, and D. T. Bergado. Properties and potential application of the selected natural fibers as limited life geotextiles. *Carbohydrate Polymers* 82 (2010): 1090–1096. doi:10.1016/j.carbpol.2010.06.036.
62. Beg, M. D. H, and K. L. Pickering. Accelerated weathering of unbleached and bleached Kraft wood fibre reinforced polypropylene composites. *Polymer Degradation and Stability* 93 (2008): 1939–1946. doi:10.1016/j.polymdegradstab.2008.06.012.
63. Omrani, E., P. L. Menezes, and P. K. Rohatgi. State of the art on tribological behavior of polymer matrix composites reinforced with natural fibers in the green materials world. *Engineering Science and Technology, an International Journal* 19 (2016): 717–736. doi:10.1016/j.jestch.2015.10.007.

64. Behera, F. K., and M. G. B. Murali. Study of the mechanical properties & erosion properties of jute-epoxy composite filled with sic particulate. *International Journal of Research in Engineering & Science* 3 (2015): 10–18.

65. Yallew, T. B., P. Kumar, and I. Singh. Sliding wear properties of jute fabric reinforced polypropylene composites. *Procedia Engineering* 97 (2014): 402–411. doi:10.1016/j.proeng.2014.12.264.

66. Wei, C., M. Zeng, X. Xiong, H. Liu, K. Luo, and T. Liu. Friction properties of sisal fiber/nano-silica reinforced phenol formaldehyde composites. *Polymer Composites* 36 (2015): 433–438.

67. Ahmed, K. S., S. S. Khalid, V. Mallinatha, and S. A. Kumar. Dry sliding wear behavior of SiC/Al$_2$O$_3$ filled jute/epoxy composites, *Mater & Design* 36 (2012): 306–315.

8

Hydrothermal Effect on the Mechanical Properties of Multiple Nanofillers Embedded Hybrid FRP Composites

8.1 Introduction

A fiber-reinforced polymer composite (FRP) is a class of materials belonging to the composite family which is usually known for its high strength-to-weight ratio and non-corrosive nature. A formal definition of composite follows from a combination of two or more phases in which the reinforcing phase is dispersed within a continuous matrix phase. The reinforcing phase may be in the form of particulates, flakes, or fibers. Out of those choices, fibers as a reinforcing phase have emerged as the most efficient candidates for several applications in structural, aerospace, automotive, and marine sectors [1]. According to a survey, it is found that the global market size of FRPs will amount to US $39,400 million by 2026 from US $31,900 million in 2019 (base year) at a compound annual growth rate (CAGR) of 2.7% by considering the key players in the industry (BASF, DuPont, Lanxess, DSM, PolyOne, SGL, etc.) [2]. One of the earliest used FRP materials was glass fiber-embedded polymer resin by a growing petrochemical industry just after World War II. Later in the 1960s and 1970s, fibers of low density having better strength and stiffness (e.g., boron, carbon, and agamid) were commercialized to fulfill the high operational requirements of aerospace/space exploration. However, the big budget was a major hindrance in creating a greater impact beyond its use in high-end applications [3]. As time progressed, there was a growing demand for these technologically advanced fiber-reinforced polymer composite materials, and their decreasing cost was favorable for their widespread use. But the performance of these materials was discouraging under various accelerated environmental conditions. Many researchers have reported the detrimental effect of long-time exposure to elevated/cryogenic temperatures, humidity, and immersion in water on the overall properties of the composite materials [4]. They may be subjected to the combined effect of variations of both temperature and moisture levels when in service. Such an effect is often named as the hydrothermal effect, which damages the mechanical properties of the polymer composites more than the individual components [5,6]. The major consequence of hydrothermal degradation appears in the form of interfacial damage between the filler and the matrix, which directly/indirectly affects the lifetime performances and structural integrity of the composites. This is because the polymers are very much prone to the hydrophilic nature of the hydroxyl groups present [7]. Such degradation is even more proliferated with temperature variations. So, it becomes of utmost importance to incorporate nanofillers (e.g., carbon nanofillers), which are hydrophobic and can reduce moisture absorption. The adverse effects of this phenomenon include weakening of bonds between the matrix and filler plasticization of the matrix phase, generation of residual stresses near interfaces, swelling of composites, and formation of cracks due to osmotic pressure. In this chapter, an attempt has been made to address the hydrothermal effect and its related issues that affect the properties, particularly, the mechanical properties of hybrid FRPs with nanofillers.

8.2 Organic-Organic Nanofillers

There has been a lot of work on various types of fillers, out of which one of the possible combinations is the hybrid of organic-organic nanofillers. Most of the agro and natural fibers along with carbon fibers come under the umbrella of organic nanofillers. The main advantages of organic fillers are their low cost and biodegradable nature (reduces environmental pollution). They also contribute to the rigidity and thermo-mechanical properties of the matrix to a great extent, exploiting their application in areas of automotive and construction industries (e.g., water storage vessels, leisure boats, desalination plants, and sports items, etc.). However, the natural FRP composites are normally hygroscopic and hydrophilic, which considerably influences their physico-mechanical properties. Specifically, the dimensional steadiness and the viscoelastic properties are major victims of the moisture content of the composites, which in turn causes their aging in the hydrothermal atmosphere. Hence, through proper choice of fiber and matrix, the performance of the composites can be improvised by the method of hybridization. By integrating carbon fibers with natural fibers (jute, hemp, abaca, and banana), numerous FRP hybrid nanocomposites have been developed and reported in the literature. While carbon/sisal hybrid composites are of particular interest in aerospace interior structures due to their low weight and optimum impact performance, flax-based hybrid composites are known for their high tensile strength (450–1500 MPa) and Young's modulus (27.6–38 GPa) [8–12]. Addition of nylon fabric to coir pith-reinforced epoxy not only increased the tensile, impact, and flexural strength of the composites, but could also address one of the major drawbacks of coir pith fillers, i.e., their rapid water uptake and retention property which causes aging of the composite panels in moist conditions [13]. At the interface, the water-filled voids may germinate cracks and microvoids due to interfacial de-bonding [13,14]. Finally, the pith swells and matrix tend to chain reorientation, leading to degradation of mechanical properties. However, the incorporation of hydrophobic nylon into chemically treated pith and epoxy hybrid composites helped in retaining the mechanical properties to a great extent, even in the presence of moisture. Actually, the chemical treatment (Figure 8.1) in the alkaline environment removes the waxy materials from the coir pith surface, improving the surface roughness, and also breaks the hydrogen bonds, promoting the activation of hydroxyl groups of cellulose units leading to less hydrophilic activity. Similarly, the aging resistance of jute/basalt fiber hybrid laminates was studied in the sandwich and intercalated stacking sequences (Figure 8.2) [15]. Those composites when exposed to hydrothermal stress and UV radiation cyclic conditions to promote accelerated aging for a long period of 84 days, alternate cycles resulted in progressive damage of the interface between the filler and matrix leading to the formation of microcracks. The photodegradation caused by UV radiation is accelerated in the presence of moisture for certain exposure time; the interface worsens and decreases the stress transmission efficiency. This kind of degradation can be postponed when basalt was used as a thicker external layer in the sandwich-like configuration. A similar response was shown by these sandwich structures to salt fog aging and was

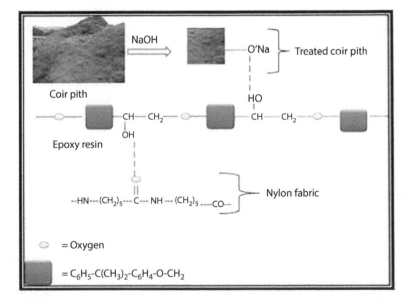

FIGURE 8.1
Hydrogen bonding interaction between chemically treated coir pith with epoxy resin and nylon 6 fabric. (From Narendar, R. et al., *Mater. Des.*, 54, 644–651, 2014.)

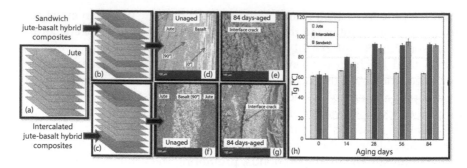

FIGURE 8.2
(a)–(c) The stacking sequences of hybrid composites; (d)–(g) SEM micrographs of unaged and 84-days-aged hybrid composites; and (h) glass transition temperature with varying aging time. (From Fiore, V. et al., *Compos. Struct.*, 160, 1319–1328, 2017.)

found to be suitable for marine applications [16]. Sergi et al. investigated the degradation of the mechanical properties of hemp/basalt hybrid-reinforced polyethylene (high density) nanocomposites after a long environmental exposure duration (water and hygrothermal aging) [17]. The hybridization of hemp with basalt remarkably decreased the water uptake nearly up to 75%, while it increased the mechanical property retention after accelerated aging, and presents a worthy replacement for glass fibers.

8.3 Organic-Inorganic Nanofillers

From the above discussion, it is quite understandable that natural fibers constitute a major portion of organic fillers, and no doubt these environmentally friendly fillers have achieved tremendous success in the auto and structural industries (such as door trim panels, engine cover, and transmission cover). But their properties are imperative in connection with low water resistance, swelling (low-dimensional stability), and vulnerable to rotting. Water not only affects the fiber, but also the matrix, due to inferior interfacial bonding causing the interaction loss between the matrix and the fiber filler. This, in turn, disturbs the mechanical integrity of the composite. The composite absorbs moisture by the volumetric diffusion process following Fick's law. The absorption of water can be reversible if it remains in microcracks/voids and can be easily removed by environmental conditioning. The problem arises when the whole process becomes irreversible; water forms hydrogen bonds/cross bonds with the long polymer chain reducing the glass transition temperature and leading to matrix plasticization. The situation becomes more critical when both water and temperature are associated. Such a problem can be reduced to a great extent

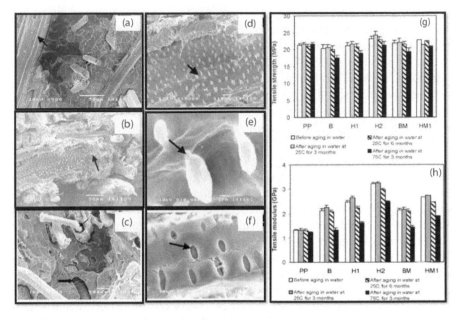

FIGURE 8.3
SEM fracture surface of bamboo-glass fiber-reinforced polypropylene hybrid composite (a) before aging; after aging for 3 months in water; (b) at 25°C; and (c) at 75°C; (d) after aging for 6 months; (e) growing polymer rods; (f) well-arranged polymer layer with holes after separation of polymer rods; and (g) and (h) tensile strength and tensile modulus before and after aging, respectively. (From Thwe, M.M. and Liao, K., *Compos. Sci. Technol.*, 63, 375–387, 2003.)

by using inorganic fillers like glass fibers in addition to organic (natural) fibers. Thwe and Liao [18] compared the hydrothermal aging resistance of injection molded bamboo/glass fiber-reinforced polymer composites with bamboo fiber-reinforced composites in terms of their mechanical properties (Figure 8.3). The samples were exposed to water at 25°C and 75°C for a prolonged period (3–6 months), and it was noticed that the bamboo/glass fiber-reinforced polymer showed better resistance to this environmental aging than bamboo fiber-reinforced polymer in respect to tensile strength and stiffness, though the extent of deterioration depends on the aging time and temperature. This can be further addressed by improving the interfacial bonding using a compatibilizer (maleic anhydride polypropylene). On the same note, jute/glass fiber-reinforced hybrid unsaturated polyester exhibited better retention of mechanical properties (tensile and flexural strength) than simply a jute-reinforced counterpart [19]. Pultruded jute and jute/glass-based composites with the fiber of high-volume fraction were kept in distilled water at room temperature for 4076 hours. The acoustic and scanning electron microscopy (SEM) observations suggest a weakening of the matrix-fiber interface due to water aging as the damage mechanism in jute fibers as fillers, which can be mitigated by incorporating the hybridization of jute by glass fibers. A combination of hemp with short glass fibers enhanced the strength of hemp-reinforced polypropylene composites and optimum mechanical properties (flexural strength ᴗ101 MPa, flexural modulus ᴗ5.5 GPa) could be achieved at 15 wt% short glass fiber [20]. The introduction of glass fibers enhanced both the water absorption and thermal stability of the composites. Aging studies on a hybrid of curaua/glass fiber with orthophalic polyester as the matrix was carried out by Silva et al. using both seawater and distilled water. One of the main advantages of curaua fiber is its better mechanical performance as compared to the vegetal fibers like sisal, jute, hemp, etc. when hybridized with glass fibers; it can be used in moisture (high%) containing environments or direct water contacts, such as reservoirs, piping, packing, and buildings [21,22]. The water absorption tests indicate better conduct of the hybrid composites as compared to the virgin natural fiber composites. Likewise, incorporation of glass fibers into *purpureum*-epoxy composites improved its tensile and flexural strength under wet conditions, although the properties degraded after durations of exposure were increased due to weakening the interfacial bond between matrix and filler [23].

Apart from natural organic fillers, there also exists synthetic organic fillers such as carbon fibers and Kevlar [24]. Carbon fibers, in particular, are a vibrant class of fillers which is available in many forms and are nowadays important from the point of view of overhead conductors for power distribution. In such applications, the overhead power lines need to be supported by those carbon fiber-based hybrid composites. In such cases, the long-term durability of the conducting composites is of utmost importance, with less maintenance over a service lifespan of multiple decades. The above concerns mainly arise from the exposure to environmental conditions including temperature, high moisture content, radiations, aggressive chemicals, etc. In this context, thermal aging studies on carbon/glass fiber hybrid composites have been investigated by Barjasteh et al. [25] to determine the oxidation kinetics and degradation mechanisms using a reaction-diffusion model. In another study, it was revealed that thermal exposure (atmospheric environment up to a year) deteriorates the flexural properties under static and fatigue loading, of the composites [26]. SEM micrographs also support the minor reduction observed in the mechanical properties; it can be attributed to thermal oxidation at intermediate aging times, while degradation at longer aging times was due to thermal aging. Recently, Jesthi and Nayak have studied the influence of the stacking sequence on the seawater diffusivity of flexural/impact/tensile strength of carbon/glass fiber hybrid composites (Figure 8.4). A reduction of 44% in seawater diffusivity was observed for [CG2CG]$_S$ hybrid composites compared to the plain glass-FRP composites. The retention of the tensile/flexural/impact strength of [CG2CG]$_S$-type hybrid composites after seawater aging were ᴗ95%, 82%, and 94%, respectively [27].

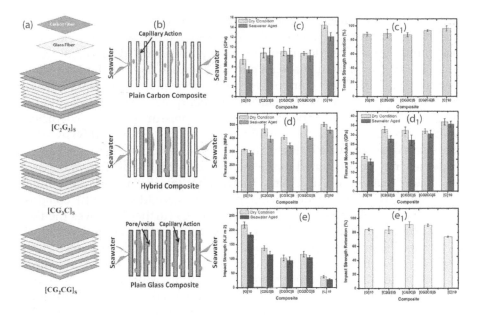

FIGURE 8.4
(a) Stacking sequence of carbon and glass fiber in hybrid composite laminates; (b) seawater absorption mechanism; (c) and (c_1) tensile modulus and tensile strength retention; (d) and (d_1) flexural stress and flexural modulus; and (e) and (e_1) impact strength and impact strength retention for different composites in dry and seawater condition. (From Jesthi, D.K. and Nayak, R.K., *Compos. Part B-Eng.*, 106980, 2019.)

8.4 Inorganic-Inorganic Nanofillers

Glass fiber-reinforced polymer (GFRP) composites constitute one of the key players (among the inorganic category) in composite industries owing to their light weight, high specific strength, corrosion resistance, and economic viability as compared to their metallic and non-metallic counterparts. All these possibilities lead to extensive use of the GFRP in various parts of aerospace, space, marine, automobile, and civil infrastructure. However, the environmental degradation caused by various factors (mentioned earlier) affects the real-time use of those composites. The mechanical properties deteriorate due to microcrack formation by uneven thermal expansion and contraction under hydrothermal conditions. Additionally, the water absorption gets accelerated with prolonged immersion because of capillary action and the hydrophilic groups in glass fibers. It is observed that the introduction of inorganic nano-/microfillers (e.g., $CaCO_3$, Al_2O_3, TiO_2, and SiO_2) reduces the detrimental effect of the hydrothermal atmosphere by improving the interfacial bond due to high specific area. Also, the incorporation of nano-sized $Al_2O_3/SiO_2/TiO_2$ lessens the water permeability and affinity, giving rise to corrosion and hydrothermal degradation resistance [28–30]. Nayak et al. made an effort to increase the inter-laminar shear strength (ILSS) of GFRP composites (with epoxy matrix) by simultaneous dispersion of Al_2O_3/TiO_2 nanoparticles [31]. Incorporation of 0.1 wt% of nano-TiO_2 (alone) into GFRP composites reduces the moisture diffusion coefficient by ~9%, enhances the residual flexural strength by ~19%, and ILSS by ~18% as compared to pure GFRP (Figure 8.5) [32]. Such an upswing in mechanical properties may be attributed to the microcrack bridging by TiO_2 nanoparticles, matrix toughening, and better matrix-fiber interfacial bonding. Similar observations were also made when nano-Al_2O_3 was added to GFRP composites [7]. Incorporation of both of these nanofillers further improves the flexural strength/modulus and ILSS in the hydrothermal environment, which is evidenced from the appearance of river line marks, highly oriented shear cusps on field emission scanning electron microscopy (FESEM) fractured surface images [31]. In another study, Nayak et al. [33] observed that when TiO_2 nanoparticles of 0.1 wt% were added to epoxy, the seawater diffusivity enhanced by 15%, whereas the flexural and ILSS increased by 15% and 23%, respectively, after seawater aging (Figure 8.6). The clay, an important inorganic filler, is commonly added to thermoplastic polymers to reduce cost and improve the mechanical performance of the composites (Young's modulus,

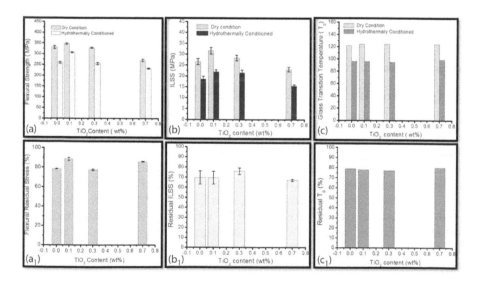

FIGURE 8.5
Effect of TiO_2 on (a) flexural strength, (a_1) residual strength, (b) ILSS in dry and hydrothermal condition, (b_1) residual ILSS after hydrothermal condition, (c) glass transition temperature in dry and hydrothermal condition, and (c_1) residual Tg after hydrothermal condition of nanocomposites. (From Nayak, R.K. et al., *Compos. Part A-Appl. Sci. Manuf.*, 90, 736–747, 2016.)

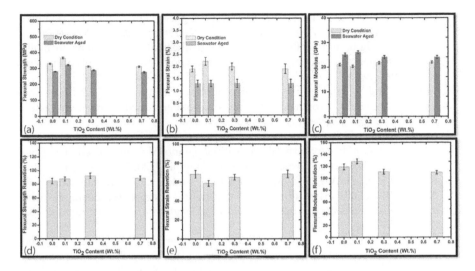

FIGURE 8.6
Effect of TiO$_2$ content on (a) flexural strength, (b) strain, (c) modulus, on retention of (d) flexural strength, (e) strain, and (f) modulus of glass fiber-reinforced polymer nanocomposites. (From Nayak, R.K. and Ray, B.C., *Arch. Civil Mech. Eng.*, 18, 1597–1607, 2018.)

heat distortion temperature). Clay in combination with other fiber fillers (hybridization) leads to synergetic effects, i.e., fracture toughness, stiffness, wear resistance, and dynamic response, etc. When Morales and White tried to study the effect of hydrothermal aging on the mechanical properties and the residual stress distribution profile of injection-molded hybrid sepiolite and talc/short glass fiber/polypropylene composites, they observed much improvement in properties [34]. A similar kind of investigation was also carried out on glass/polypropylene/epoxy composites [35].

References

1. Masuelli, M. A. Introduction of fibre-reinforced polymers – Polymers and composites: Concepts, properties and processes. In M. Masuelli (Ed.), *Fiber Reinforced Polymers-The Technology Applied for Concrete Repair*. 2013. IntechOpen, Rijeka, Croatia.
2. https://www.marketwatch.com/press-release/fiber-reinforced-plastics-frp-market-report-by-sales-and-revenue-analysis-2019-to-2026-2019-05-15?mod=mw_quote_news.
3. Bakis, C. E., L. C. Bank, V. Brown, E. Cosenza, J. F. Davalos, J. J. Lesko, A. Machida, S. H. Rizkalla, and T. C. Triantafillou. Fiber-reinforced polymer composites for construction—State-of-the-art review. *Journal of Composites for Construction* 6.2 (2002): 73–87.

4. DiBenedetto, A. T. Tailoring of interfaces in glass fiber reinforced polymer composites: A review. *Materials Science and Engineering: A* 302.1 (2001): 74–82.
5. Lin, Q., X. Zhou, and G. Dai. Effect of hydrothermal environment on moisture absorption and mechanical properties of wood flour–filled polypropylene composites. *Journal of Applied Polymer Science* 85.14 (2002): 2824–2832.
6. Saha, S., and S. Bal. Long term hydrothermal effect on the mechanical and thermo-mechanical properties of carbon nanofiber doped epoxy composites. *Journal of Polymer Engineering* 38.3 (2018): 251–261.
7. Nayak, R. K., and B. C. Ray. Water absorption, residual mechanical and thermal properties of hydrothermally conditioned nano-Al_2O_3 enhanced glass fiber reinforced polymer composites. *Polymer Bulletin* 74.10 (2017): 4175–4194.
8. Khanam, P. N., H. A. Khalil, M. Jawaid, G. R. Reddy, C. S. Narayana, and S. V. Naidu. 2010. Sisal/carbon fibre reinforced hybrid composites: tensile, flexural and chemical resistance properties. *Journal of Polymers and the Environment* 18.4 (2010): 727–733.
9. George, J., and J. I. I. Verpoest. Mechanical properties of flax fibre reinforced epoxy composites. *Die Angewandte Makromolekulare Chemie* 272.1 (1999): 41–45.
10. Sarasini, F., J. Tirillò, S. D'Altilia, T. Valente, C. Santulli, F. Touchard, L. Chocinski-Arnault, D. Mellier, L. Lampani, and P. Gaudenzi. 2016. Damage tolerance of carbon/flax hybrid composites subjected to low velocity impact. *Composites Part B: Engineering* 91 (2016): 144–153.
11. Živković, I., C. Fragassa, A. Pavlović, and T. Brugo. Influence of moisture absorption on the impact properties of flax, basalt and hybrid flax/basalt fiber reinforced green composites. *Composites Part B: Engineering* 111 (2017): 148–164.
12. Almansour, F. A., H. N. Dhakal, and Z. Y. Zhang. Effect of water absorption on Mode I interlaminar fracture toughness of flax/basalt reinforced vinyl ester hybrid composites. *Composite Structures* 168 (2017): 813–825.
13. Narendar, R., K. P. Dasan, and M. Nair. Development of coir pith/nylon fabric/ epoxy hybrid composites: Mechanical and ageing studies. *Materials & Design (1980–2015)* 54 (2014): 644–651.
14. Khalil, H. A., M. Jawaid, and A. A. Bakar. Woven hybrid composites: Water absorption and thickness swelling behaviours. *BioResources* 6.2 (2011): 1043–1052.
15. Fiore, V., T. Scalici, D. Badagliacco, D. Enea, G. Alaimo, and A. Valenza. Aging resistance of bio-epoxy jute-basalt hybrid composites as novel multilayer structures for cladding. *Composite Structures* 160 (2017): 1319–1328.
16. Fiore, V., T. Scalici, F. Sarasini, J. Tirillò, and L. Calabrese. Salt-fog spray aging of jute-basalt reinforced hybrid structures: Flexural and low velocity impact response. *Composites Part B: Engineering* 116 (2017): 99–112.
17. Sergi, C., J. Tirillò, M. C. Seghini, F. Sarasini, V. Fiore, and T. Scalici. Durability of basalt/hemp hybrid thermoplastic composites. *Polymers* 11.4 (2019): 603.
18. Thwe, M. M., and K. Liao. Durability of bamboo-glass fiber reinforced polymer matrix hybrid composites. *Composites Science and Technology* 63.3–4 (2003): 375–387.
19. Akil, H. M., C. Santulli, F. Sarasini, J. Tirillò, and T. Valente. Environmental effects on the mechanical behaviour of pultruded jute/glass fibre-reinforced polyester hybrid composites. *Composites Science and Technology* 94 (2014): 62–70.
20. Panthapulakkal, S., and M. Sain. Injection-molded short hemp fiber/glass fiber-reinforced polypropylene hybrid composites—Mechanical, water absorption and thermal properties. *Journal of Applied Polymer Science* 103.4 (2007): 2432–2441.

21. Silva, R. V., E. M. F. Aquino, L. P. S. Rodrigues, and A. R. F. Barros. Curaua/ glass hybrid composite: The effect of water aging on the mechanical properties. *Journal of Reinforced Plastics and Composites* 28.15 (2009): 1857–1868.
22. Rodrigues, L. P. S., R. V. Silva, and E. M. F. Aquino. Effect of accelerated environmental aging on mechanical behavior of curaua/glass hybrid composite. *Journal of Composite Materials* 46.17 (2012): 2055–2064.
23. Ridzuan, M. J. M., M. A. Majid, M. Afendi, K. Azduwin, N. A. M. Amin, J. M. Zahri, and A. G. Gibson. Moisture absorption and mechanical degradation of hybrid Pennisetum purpureum/glass–epoxy composites. *Composite Structures*, 141 (2016): 110–116.
24. Felipe, R. C. T. S., R. N. B. Felipe, A. C. M. C. Batista, and E. M. F. Aquino. Influence of environmental aging in two polymer-reinforced composites using different hybridization methods: Glass/Kevlar fiber hybrid strands and in the weft and warp alternating Kevlar and glass fiber strands. *Composites Part B: Engineering* 174 (2019): 106994.
25. Barjasteh, E., E. J. Bosze, Y. I. Tsai, and S. R. Nutt. Thermal aging of fiber-glass/carbon-fiber hybrid composites. *Composites Part A: Applied Science and Manufacturing* 40.12 (2009): 2038–2045.
26. Burks, B., and M. Kumosa. The effects of atmospheric aging on a hybrid polymer matrix composite. *Composites Science and Technology*, 72.15 (2012): 1803–1811.
27. Jesthi, D. K., and R. K. Nayak. Evaluation of mechanical properties and morphology of seawater aged carbon and glass fiber reinforced polymer hybrid composites. *Composites Part B: Engineering* 174 (2019): 106980.
28. Golru, S. S., M. M. Attar, and B. Ramezanzadeh. Studying the influence of nano-Al_2O_3 particles on the corrosion performance and hydrolytic degradation resistance of an epoxy/polyamide coating on AA-1050. *Progress in Organic Coatings* 77.9 (2014): 1391–1399.
29. Zhao, H. and R. K. Li. Effect of water absorption on the mechanical and dielectric properties of nano-alumina filled epoxy nanocomposites. *Composites Part A: Applied Science and Manufacturing* 39.4 (2008): 602–611.
30. Shi, H., F. Liu, L. Yang, and E. Han. Characterization of protective performance of epoxy reinforced with nanometer-sized TiO_2 and SiO_2. *Progress in Organic Coatings* 62.4 (2008): 359–368.
31. Nayak, R. K., and B. C. Ray. Retention of mechanical and thermal properties of hydrothermal aged glass fiber-reinforced polymer nanocomposites. *Polymer-Plastics Technology and Engineering* 57.16 (2018): 1676–1686.
32. Nayak, R. K., K. K. Mahato, and B. C. Ray. Water absorption behavior, mechanical and thermal properties of nano TiO_2 enhanced glass fiber reinforced polymer composites. *Composites Part A: Applied Science and Manufacturing* 90 (2016): 736–747.
33. Nayak, R. K., and B. C. Ray. Influence of seawater absorption on retention of mechanical properties of nano-TiO_2 embedded glass fiber reinforced epoxy polymer matrix composites. *Archives of Civil and Mechanical Engineering* 18.4 (2018): 1597–1607.
34. Morales, E., and J. R. White. Effect of ageing on the mechanical properties and the residual stress distribution of hybrid clay–glass fibre–polypropylene injection mouldings. *Journal of Materials Science* 44.17 (2009): 4734–4742.
35. Abd El-Baky, M. A. Experimental investigation on impact performance of glass–polypropylene hybrid composites: Effect of water aging. *Journal of Thermoplastic Composite Materials*, 32.5 (2019): 657–672.

9

Future Prospective and Challenges

The need to surpass the limitations of conventional materials has fueled a large increase in engineering applications of cutting-edge fiber reinforced polymer (FRP) composite materials in recent years. The main industries include aerospace and defense [1], marine, automotive, oil and gas industries, and construction. FRP composites form a very attractive set of materials for use in engineering applications due to their highly favorable material properties, which include the light weight that imparts them high strength-to-weight and stiffness-to-weight ratios, corrosion resistance, and a large number of variables that can be tuned according to requirements. These materials are at the leading edge of materials technology today, with performance and costs appropriate to both ultra-demanding applications and utilitarian components in everyday life. Often touted as the material of the future, the extensive acceptance of these materials arises from their structural tailorability, light weight, and cost-effective manufacturing techniques.

Fiber-reinforced composites are mainly axial particulates implanted in suitable matrices. The reinforcement of fiber upon a polymeric matrix is found to bring about significant advancements in the mechanical behavior of the polymeric host with the added advantages of a light weight, high strength-to-weight ratio, excellent weathering stabilities, and enhanced dimensional stabilities, in addition to low maintenance costs and tailor-made material behaviors. However, to obtain tailor-made properties that suit specified application requirements, various types of fibers with varying polymers have been tried, with fibers contributing toward the betterment of mechanical, tribological, thermal, and water sorption behavior of the resulting composites. Figure 9.1 illustrates timeline for development of composite materials. One could expect Herculean results when the fibers are near-to-infinite in length, isotropic, and are inserted unidirectional. Conversely, the greater the anisotropy and shorter the length of the fibers, the lower would be their overall mechanical performance. The observed large leeway regarding length, direction, and type of fiber widen the application window of FRP composites. Further, the composite can be downright made-to-measure to suit the exact mechanical requirements for any particular project, which in turn supports far better effectiveness toward the end applications. The most common fibers exercised in recent times are glass fibers, carbon fibers, aramid fibers, and natural fibers, next to which is the nylon- and polyester fibers. All of these reinforcements have pretty low densities. The excellent strength-to-weight and stiffness-to-weight ratios

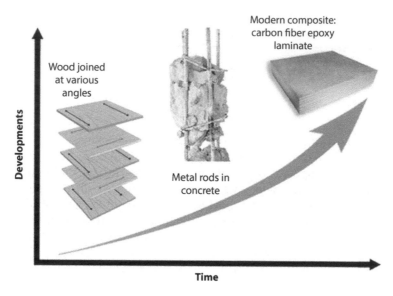

FIGURE 9.1
Timeline for the development of composite materials. (From https://www.taylorfrancis.com/books/e/9780429175497.)

of composites are therefore credited to low fiber densities. Nevertheless, all fibers behave quasi-elastically until breakage, with carbon fibers much stiffer and lighter than glass fibers. This, in turn, accounts for the increased preference for them in many high-performance applications.

Fibers are generally not used alone. They need to be impregnated in a matrix that acts to transfer the load of the reinforcements (fibers). The matrix also protects the fibers from environmental attack and abrasion. The matrix, of course, reduces the properties to some extent, but even so, the very high definite (weight-adjusted) properties are available from these materials. Glass and metals are available as matrix materials, but these are largely restricted to research and development (R&D) laboratories and are currently very expensive. Polymers are much more commonly used, with unsaturated styrene-hardened polyesters having the majority of the low-to-medium performance applications, and epoxy or more sophisticated thermosets having the higher end of the market. Epoxy has been commonly used as a matrix material in structural load-bearing FRP components due to its superior strength, stiffness, and thermal and chemical resistance. Thermoplastic matrix composites are gaining attention, increasingly, with processing complications being possibly their principal restriction.

From airplanes to surfboards to cylinders, fiber-reinforced polymer composites are all around us now. But these heterogeneous materials combining the best aspects of dissimilar constituents have been used by nature for millions of years—trees employ cellulosic fibrous components to reinforce a lignin matrix and arrange the strong fibers in just the correct direction to withstand loads from wind and other environmental sources. We can get a large combination of elastic modulus and strength from an FRP, which is very much similar to or even better than many conventional metallic materials (Table 9.1). In Table 9.1, the strength and modulus of unidirectional composites are taken in the direction of fibers. It is also seen that many of the FRP composites have excellent fatigue strength and are tolerant of fatigue damage.

Imitating nature, the ancient societies used this methodology as well. The earliest of man-made composites, made of mud and straw, date back some 6000 years to Egypt. Ancient Mesopotamians developed plywood at around 3400 BC, gluing and joining wood at various angles for improved durability and strength. The Book of Exodus speaks of adding metal rods or wires to the concrete to increase its tensile strength. Fiber-reinforced polymer composites are among the newest composites, dating back to around the dawn of the twentieth century. The modern use of FRPs began in the years near World War II, with early applications being rocket motor cases and radomes using glass fibers. The Chevrolet Corvette of the 1950s had a fiberglass body (they still do), which wasn't always easy to repair after a collision, but did offer an escape from the rusting that afflicts most car bodies. Growth of the market for composites has been very good overall since the 1960s, averaging around 15% per year compared with 11% for plastics, 6½% for chemicals, and 3½% for the gross domestic product (GDP).

The type and quantity of the reinforcement are responsible for determining the final properties of the composite. It is the reinforcement which provides strength and stiffness. It may be a fiber or a particulate. The matrix is crucial for transferring the load to fibers through shear loading. It also maintains the proper orientation and spacing of fibers, thereby protecting them from abrasion. The key feature behind the design and manufacture of high-performance composites with specific functional requirements is the good changeability potential of composites. The physical property tensors of a particular material can be customized over a wide range by varying its structural parameters like composition, symmetry, connectivity, and periodicity.

In terms of design, the FRP composite structures are much more complex than the metal structures. But the anisotropy of the FRP composites allows tailoring the properties according to design requirements. This is an advantage of the FRP composites that can be effectively utilized in a number of ways, reinforcing a structure selectively in the direction where stress is more and where the stiffness is needed to increase. The fabrication of curved panels with no secondary forming operation or negligible thermal expansion coefficient structures can be produced. A very unique characteristic of FRP composites, which provides it another degree of design flexibility which cannot be seen even in metals, is the skin material and a lightweight core, which is in the form of an aluminum honeycomb, plastic, or metal foam that can

TABLE 9.1

Tensile Properties of Some Metallic and Structural Composites

Materials	Density g/cm³	Modulus GPa	Tensile Strength MPa	Yield Strength MPa	Specific Modulus, 10⁶ m	Specific Strength, 10³ m
SAE 1010 steel (cold-worked)	7.87	207	365	303	2.68	4.72
AISI 4340 steel (quenched and tempered)	7.87	207	1722	1515	2.68	22.3
6061-T6 aluminum alloy	2.70	68.9	310	275	2.60	11.7
7178-T6 aluminum alloy	2.70	68.9	606	537	2.60	22.9
Ti-6A1-4V titanium alloy (aged)	4.43	110	1171	1068	2.53	26.9
17-7 PH stainless steel (aged)	7.87	196	1619	1515	2.54	21.0
INCO 718 nickel alloy (aged)	8.2	207	1399	1247	2.57	17.4
High-strength carbon fiber—epoxy matrix (unidirectional)	1.55	137.8	1550	–	9.06	101.9
High-modulus carbon fiber—epoxy matrix (unidirectional)	1.63	215	1240	–	13.44	77.5
E-glass fiber—epoxy matrix (unidirectional)	1.85	39.3	965	–	2.16	53.2
Kevlar49 fiber—epoxy matrix (unidirectional)	1.38	75.8	1378	–	5.60	101.8
Boron fiber-6061 Al alloy matrix (annealed)	2.35	220	1109	–	9.54	48.1
carbon fiber—epoxy matrix (quasi-isotropic)	1.55	45.5	579	–	2.99	38
Sheet-molding compound SMC (isotropic)	1.87	15.8	164	–	0.86	8.9

Source: FRP Defence Products—Suvarna Fibrotech Pvt. Ltd. n.d. http://www.suvarnafrpproducts.com/index.php/category-blog/2015-01-08-04-08-35/defence-products (Accessed June 15, 2019).

be used to build a sandwich beam, plate, or shell structure which can give a very high stiffness without any increase in weight.

FRPs in automotive applications can provide passengers with great comfort. This will result in better vibration absorption in the material and will also reduce the vibration and noise to the nearby structures. Fiberglass composites improve the strength and performance of the present highway infrastructure located in the severe environments that could be the chief reason why the original structures weakened [2]. There are many advantages of FRP composites, one of which is its non-corroding nature. Still they are prone to environmental damages due to the absorption of moistures or chemicals, which can lead to a change in the dimension or even internal stress in the material.

The hand lay-up technique is a simple and low-cost molding method where a resin coat is first applied to the mold, and then, manually, a fiber reinforcement is placed on it. Rollers are used to cohere the laminate and remove any entrapped air. Subsequently, more fiber layers are added to make the laminate of the desired thickness. In automated spray layup, using a handheld gun, the fiber is chopped in and fed into a spray of catalyzed resin. But the major disadvantage is that excessively heavy and resin-rich laminates are produced. The incorporation of only short fibers severely restricts the mechanical properties of the prepared laminate. The applications consist of lightly laden structural panels, simple enclosures, e.g., bathtubs, caravan bodies, some small dinghies, and shower trays. Figure 9.2 shows Bus stand made up of FRP [3]. In the process of resin infusion, the resin is poured into the voids of an evacuated stack of porous materials. After solidification of the resin, it binds all the materials into a stiff composite. A vital part of this process is the evacuation of air from the porous material before adding the resin.

The bulk molding process incorporates a bulk molding compound, which is a ready-to-mold, glass fiber-reinforced thermoset polymer material. It is manufactured at room temperature by mixing strands of chopped glass fibers with initiator styrene and filler in a thermoset resin. The mixture is

FIGURE 9.2
Bus stand made up of FRP. (From https://www.taylorfrancis.com /books/e/9780429175497.)

stored at low temperatures so that curing is slowed down. Sheet molding can be performed by compression or injection molding. The compounds are a combination of long chopped glass strands, thermosetting resin, and mineral fillers. The pultrusion process produces composites with high fiber volume fraction and high strength-to-weight ratio. The reinforcement is pulled into the infeed area, and then impregnated with the resin matrix. It is then cured and allowed to cool. After that, it is clamped and pulled by the reciprocating puller units.

There is so much variation in the commercial and industrial applications of fiber-reinforced polymer composites that it is almost impossible to list them all. Some of the major structural applications are discussed below. Military and commercial aircraft see major structural applications for FRP composites. FRP composites are in this field, as they demand weight reduction, which is critical for higher speeds and increased payloads. In 1969, FRPs witnessed major growth in the aircraft industry because of the production application of boron fiber-reinforced epoxy skins for F-14 horizontal stabilizers. In the 1970s, carbon fibers were introduced, thereby making carbon fiber-reinforced epoxy the primary material in many wings, fuselage, and empennage components.

Corrosion is the major problem which is faced with the use of steel or aluminum alloys, along with the environmental degradation suffered by wood. So, in the initial stages, FRP composite applications were henceforth directed to overcome these drawbacks. Additionally, the topside weight of the ships could be reduced with the use of composites. Ship radomes, as well as the sonar domes of submarines, benefited from the high acoustic transparency of composites. The FRP patrol boat is increasingly becoming popular mainly due to light weight composites and excellent corrosion resistance, which reduce maintenance costs. This results in better fuel economy and speed. It is estimated that the FRP composite patrol boats are generally approximately 10% lighter than an aluminum boat and over 35% lighter than a steel boat of the same size. The high cost of carbon fiber composites limits their usage in naval vessels. Composite made propulsion shafts are predicted to be 18%–25% lighter than steel shafts of the same size, along with the reduction in life cycle cost by a minimum of 25% because of fewer issues related to corrosion and fatigue.

Due to the high strength-to-weight and stiffness-to-weight ratios, as well as light weight and corrosion resistance, FRP composites have potential applications in civil engineering. The renewal of constructed facilities and infrastructure, such as buildings, bridges, pipelines, etc. are of the upmost importance. Due to the changeable performance characteristics, ease of application, and low life cycle costs, the use of FRP composites have been increased in the rehabilitation of concrete structures and development of a new light weight structural concept. In the automotive industry, the classification of FRP composites on the basis of applications can be made into three groups: body components, chassis components, and engine components.

Exterior body components, such as body panels, hood, bumpers, and door panels, require high stiffness, damage tolerance, dent resistance, as well as a first class surface finish for appearance. E-glass fiber-reinforced sheet molding compound (SMC) composites are suitable composite materials for this application. The tooling cost for compression molding SMC parts can be 40%–60% lower than that for stamping steel parts. An example of parts integration can be found in radiator supports in which SMC are used as a substitution for low carbon steel. With as much as 80% weight reduction, uni-leaf E-glass fiber-reinforced epoxy springs have been used to replace multi-leaf steel springs.

Composites are improving the design process and end products across industries from aerospace to renewable energy. Each year, composites carry on to replace traditional materials like aluminum and steel. As the costs of composites come down and design flexibility improves, FRP composites like carbon fiber and fiberglass open up novel design prospects for engineers. Perhaps the biggest advantage of composites is their high strength-to-weight ratio. They have high stiffness, strength, and toughness, often comparable with structural metal alloys. Further, they typically deliver these properties at significantly less weight than metals: their strength and modulus per unit weight is nearly five times that of aluminum and steel. High-end auto engineers use composites to decrease vehicle weight by as much as 60%, while improving crash safety; multilayer composite laminates absorb more energy than traditional single-layer steel. In weight-critical devices such as airplanes or spacecraft, these weight savings are a compelling advantage.

Composites do not rust as do many ferrous alloys, and resistance to this common form of environmental degradation may offer better life cycle cost even if the original structure is initially costlier. Their fracture toughness is less than metals, but more than most polymers. Composites have high dimensional stability, which permits them to preserve their shape, in conditions of varying temperature and humidity. This makes them a widespread material for outdoor structures like wind turbine blades. Engineers choose composites over other traditional materials to diminish maintenance costs and guarantee long-term stability. These are major benefits for structures and systems built to last a long time. Composites can be made anisotropic, i.e., have different properties in different directions. This property of anisotropy can be used to design a more effective structure. In various structures, the stresses are also dissimilar in different directions; for example, in closed-end pressure vessels—such as a rocket motor case—the circumferential stresses are twice the axial stresses. Utilizing composites, such a vessel can be made twice as strong in the circumferential direction as in the axial.

Various structures are subject to fatigue loading, in which the internal stresses fluctuate with time. Axles on rolling stock are examples; here, the stresses vary sinusoidally from tension to compression as the axle turns. These fatigue stresses can eventually lead to failure, even when the maximum stress is much less than the failure strength of the material as measured in a static tension test. Composites possess excellent fatigue resistance when compared with metal alloys, and often display evidence of accumulating fatigue damage, which enables detection of damage and replacement of a part before catastrophic failure occurs. Materials can exhibit damping, in which a certain fraction of the mechanical strain energy deposited in the material by a loading cycle is dissipated as heat. This can be advantageous, for instance, in controlling mechanically induced vibrations. Composites generally offer relatively high levels of damping, and furthermore, the damping can often be tailored to the desired levels by suitable formulation and processing. Composites show excellence in applications involving sliding friction, with tribological wear properties coming close to those of lubricated steel. But as with any other material, composites are not perfect for all applications, and the designer needs to be aware of their drawbacks as well as their advantages. Not all applications are weight-critical. If lightweight properties are not relevant, steel and other traditional materials live up to the task at hand at lower costs.

Anisotropy and other "special" features are helpful in that they offer a great deal of flexibility in design, however, they complicate the design. The recognized tools of stress analysis used in the isotropic linear elastic design must be extended to include anisotropy, for instance, and not all designers are comfortable with these more advanced tools. Even after numerous years of touting composites as the "material of the future," economies of scale are still not well established. As a result, composites are practically always more expensive than traditional materials, so the designers have to look to composites' numerous advantages to counterbalance the additional cost. During the energy crisis period of the 1970s, automobile manufacturers were so anxious to reduce vehicle weight that they were willing to pay a premium for composites and their weight advantages. But as worry about energy efficiency diminished, the industry gradually returned to a strict lowest-cost approach in selecting materials. Hence, the market for composites in automobiles returned to a more modest rate of growth.

The key driving factors for the global acceptance of FRP composite materials as a trending structural material are:

- High modulus-to-weight ratio and strength-to-weight ratio
- Wide range of selection of raw materials as per the application requirements
- Potentiality and intrinsic resistance to weather and the corrosive upshots of salt air and seawater
- Improved fatigue resistance
- Rapid installation at the site for footbridges, platforms for oil and gas industry
- Low maintenance
- Tailorability

The immediate use of fibrous polymeric materials may be in water/sewage pipelines, electric poles, portable cabins, and toilets as a lot of modernization projects are being executed in various cities in a Smart City Mission. The key driving factors for using FRPs in these applications are their light weight, high strength, corrosion resistance, and also their durability. Polymer composites from renewable resources have also increased attention over the past few years due to environmental matters and fast sapping of unadventurous energy resources. The relocation of catalyst remnants, polymer additives, unreacted monomers, and polymerization solvents of low molecular mass fractions from the synthetic composites into the packaged food materials with consequent toxic health implications to the consumers have been stated by plentiful examinations. Due to its low toxicity, glass fiber has been accepted since its discovery. Figure 9.2 shows the bus stand made up of FRP.

In plastic waste disposal and recycling, FR plastic as a subclass of plastics is accountable for a number of disputes plus disquiets. FRPs will gain environmental sensitivity with the dawn of new more ecologically responsive matrices, such as bioplastics and UV-degradable plastics. Nowadays, the demand for both lightweight and durable materials has been the core driving vigor for the fast growth of polymer-based composites. Glass fiber reinforced polymer composites (GFRPs) involve comparatively lower energy during the manufacturing process as compared with metallic counterparts; they are more durable and have much lower thermal conductivity. The proposed use of advanced fibrous polymeric composites in the future may be its usage in combination with the more conventional materials like steel and aluminum.

FRPs can be functional to reinforce the bridges, beams, columns, and slabs of buildings. It is likely conceivable to upsurge the strength of structural members even after they have been harshly dented due to several loading conditions. The dented-reinforced concrete members would first involve the overhaul of the member by taking out the loose debris and filling in cracks and cavities with epoxy resin or mortar in such cases. For seismic strengthening of the reinforced concrete members, fiber-reinforced polymer composites can be used, as unadventurous materials pose impediments, such as the handling of heavy steel and the peril of corrosion. Also, visually examining the condition of a concrete member following a seismic event is impossible.

The driving force for using FRPs is their dimensional stability, better seismic resistance, high stiffness, and strength-to-weight ratio, the sturdiness and mechanical physiognomies of FRPs can be tailored, FRP composites guard the inner reinforcement against rust, as they can efficiently endure the severe environment, and it is informal to yield, handle, and fit FRP wraps without any hefty equipment [4]. Fiber-reinforced phenolic resins are used in a varied array of solicitations comprising electronics, aerospace, rail, mass transit, offshore water pipe systems, and ballistics and mine ventilation. The driving factor for using phenolic resin is its high-temperature solicitations, where parts must encounter smoke emission, fire safety, and combustion and toxicity requirements. They have admirable heat resistance, chemical resistance, and flame retardance and electrical non-conductivity features [5].

FRPs can be used for high-performance structural components in space applications, particularly for the space shuttle's launch vehicles. The implementation of nano-composites will be the future trend, as it requires a massive amount of investment. Therefore, before investing in the field of nanotechnology, we have to execute immediate and medium-term projects efficiently. Grouping of nanotechnology with fibrous polymeric composites can be an encouraging tool to reach the mark of environmental sustainability. In a marine environment, solicitations, where corrosion resistance and light weight are their prime advantages compared to metallic structures, FRP materials offer fabulous potential. Thus, attempts plus policies may be taken to prioritize the application of FRP composites in various naval applications. Though FRP materials have a lot of potential to replace the conventional materials, they are not succeeding at the expected rate, which can be addressed by only carrying out proper research, and the current trends of research are as follows:

- Establishing the criteria for the failure of FRP composites in various applications
- Novel design and manufacturing technologies for FRP composite structures and components
- Testing and characterization of FRP components aimed at engineering uses
- Long-term performance and durability of composites, systems, and structural elements
- Fatigue performance of composites, systems, and structural elements
- FRPs in nanoengineering applications

Potential applications of FRPs include highway structures, Prodeck bridge systems, auto skyways, utility poles, pipes, wind energy, blast protection of structures, decking for navy and sea basing, army bridging, air force towers, and some critical applications like cryotanks, impact plates and wear plates, sieves for industrial applications, industrial cooling towers and cooling fan blades, bridge decks, stringer beams, abutment panels, rebars, dowel bars, post poles, guardrail systems, sound barriers, drainage systems (pipe, culvert), automotive bodies, suspension systems for heavy vehicles, etc.

Presently, various modes of modifications for improving the properties, especially the out of the plane strength of the FRP composites, are being tested. Matrix modification, starting off with the addition of ceramic-based nanofillers and presently with carbon-based nanofillers carbon nano Fiber [CNF], carbon nano Tube [CNT], , graphene) is gaining ground as an effective process to increase properties of FRP composites, at various temperatures.

On the other hand, modification of the fiber promises improvements likewise. This approach focuses on coating the surface of the fiber in order to enhance the fiber-matrix interfacial properties. Fiber surfaces with higher surface area, wettability, and an affinity for a matrix warrant superior stress transfer to the reinforcements. Sizing, electrochemical method, plasma treatment, and hybrid fibers are promising modification methods. Out of these, electrophoretic deposition of nanofillers on conductive fiber like carbon fibers is an interesting option. The electrophoretic deposition (EPD) technique has been extensively adopted for surface alterations to develop nanoparticle-reinforced materials. EPD stands apart from other techniques because of several advantages like ease of controlling the film thickness, better homogeneity of surface, and high rate of deposition. Z-pinning, a process where reinforcements are inserted along the Z direction of continuous long fiber composites dramatically improves resistance to delamination. The following challenges can be highlighted in contrast to the development of FRP industries:

- Availability of quality raw materials
- The rising cost of raw materials and high price sensitivity of the market
- Lack of skilled labor
- Lack of awareness in the industrial world
- Quality assurance and standardization are further issues which concern manufacturers
- Lack of concrete regulatory bodies/framework
- A finite number of scientific publications
- Lack of recycling policy of FRP waste and end of life products
- The necessity to develop quality consciousness among small-scale FRP/GFRP composites manufacturers
- New products and applications need to be developed at a faster rate
- Moderate implementation of the automated FRP fabrication process

The first and the most critical problem the FRP industry is facing is the higher cost of raw material; this adds up in the price of the product, which adversely affects the cost competitiveness of FRP products in contrast to products manufactured from conventional materials. Raw materials manufacturing industries need to be set up, including producers of glass fiber, specialty resins and chemicals, and styrene monomer. Engineers and the general public should be made aware of the superiority of FRPs over conventional materials and have their trust built up in it. Several players are involved in the production and development of FRP composites, however, the industrialization of FRPs requires human resources training, design, mold design, and mold making, these services should be taught and nurtured at different workshops and institutes. Fabrication processes should be automated to reduce the need for skilled labor and to improve the quality, also to boost up the production rate. There is a need to build a "closed loop composite system" incorporating academia, suppliers, and industry. Composite labs should be set up within the educational institutes, this will build up the employable human resources ready for the composite sectors. Hybrid, nanotechnology, and automobile, these areas may be explored depending on raw materials, fabrication resources, infrastructure, and quality assurance tools. Already the sports sector is providing a vast market for FRPs. Hence, sports equipment with a lower price will attract everyone's attention. Even the big names in the industry should start exploring FRP composites in many sectors like aviation, defense, shipping, mass transportation, telecommunication, infrastructure and disaster management, oil and gas, power and renewable energy, industrial and municipal piping, electrical and electronics, and automobiles. In the architecture sector, FRPs are excellent materials for domes, skylights, and doors, and a few industries already have a strong foothold in these applications.

Mostly carbon and glass fiber are used to make the sports equipment. The demand for durable materials for rackets, hoverboards, and skis will drive the growth to the highest peak. Products that increase the strength and drive in golf clubs and hockey sticks, which decrease the overall weight, will increase the carbon fiber demand. Europe and North America are the major markets for the products. The regions, together, accounted for around 60% of the total industry share. North America represents an important economic zone for the industry, with the presence of major players and significant growth in the end-user industries. The European industry is growing at a steady rate. The growth of the industry in the overall region is mainly due to the rising demand in the construction and transport application sectors. The Asian Pacific area is forecasted to be the fastest emerging market. The FRPs market in India and China is experiencing an above-average growth owing to the rising construction and automotive industries in the region.

References

1. FRP Defence Products—Suvarna Fibrotech Pvt. Ltd. n.d. http://www.suvarnafrpproducts.com/index.php/category-blog/2015-01-08-04-08-35/defence-products (Accessed June 15, 2019).
2. The Application of FRP Composites in Highway Infrastructure|TUF-BAR. n.d. https://www.tuf-bar.com/the-application-of-frp-composites-in-highway-infrastructure/ (Accessed June 15, 2019).
3. Mallick P. K. Composites Engineering Handbook. Taylor Francis & Group. n.d. https://www.taylorfrancis.com/books/e/9780429175497 (Accessed June 15, 2019).

4. Oliveto G., and Marletta M. Seismic retrofitting of reinforced concrete buildings using traditional and innovative techniques. *ISET Journal of Earthquake Technology* 42.2–3 (2005): 21–46.
5. Brydson J. A. 23—Phenolic Resins. In J. A. Brydson, Ed., *Plastics Materials*, 7th ed., Butterworth-Heinemann, Oxford, 1999: pp. 635–667. doi:10.1016/B978-075064132-6/50064-4.

Index

Note: Page numbers in italic and bold refer to figures and tables, respectively.

Milton Keynes UK
Ingram Content Group UK Ltd.
UKHW031148141024
449569UK00024B/972

9 781032 240855